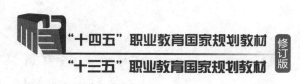

"十四五"职业教育国家规划教材 修订版
"十三五"职业教育国家规划教材

 中国电力教育协会职业院校自动化类专业精品教材

电气自动化专业英语

第三版

主　编　王　琳　汤慧芹
副主编　夏　怡　吴　梅
编　写　徐志成
主　审　陈雪丽

中国电力出版社
CHINA ELECTRIC POWER PRESS

内 容 提 要

本书连续入选"十三五""十四五"职业教育国家规划教材。

本书分为 5 章，共计 25 课。第 1 章电工电子技术基础，主要介绍电气元件与网络、网络分析、二极管、运算放大器、数字逻辑门、触发器等；第 2 章电机与电机控制，主要介绍直流电动机、交流异步电动机、电动机控制电路、脉宽调制理论、变频器等；第 3 章工业计算机控制技术，主要介绍 PLC、单片机、现场总线等；第 4 章供电技术，主要介绍火电站、太阳能、输配电、交流和直流输电系统的区别、智能电网等；第 5 章机器人技术，主要介绍机器人类型、控制方法及其嵌入式系统和视觉系统等。本书所有课文均精心选自国外相关网站和教材，具有专业性和实用性强、难度适宜等特点，有助于培养学生阅读电气类英文资料的能力。

为便于学生自学和教师教学，本书配套丰富的数字资源，如在线测试题、译文、课件等。本书为一书一码新形态教材，购买正版图书，可免费获取资源（免费码贴在封二）。

本书可作为职业教育电气自动化技术等相关专业的教材，也可供成人教育相关专业使用及专业英语爱好者学习参考。

图书在版编目（CIP）数据

电气自动化专业英语/王琳，汤慧芹主编. -- 3 版. -- 北京：中国电力出版社，2025. 2.
ISBN 978-7-5198-9406-1
Ⅰ. TM
中国国家版本馆 CIP 数据核字第 2024MD2197 号

出版发行：中国电力出版社
地　　址：北京市东城区北京站西街 19 号（邮政编码 100005）
网　　址：http://www.cepp.sgcc.com.cn
责任编辑：乔莉（010-63412535）
责任校对：黄　蓓　常燕昆
装帧设计：赵姗姗
责任印制：吴　迪

印　　刷：廊坊市文峰档案印务有限公司
版　　次：2009 年 12 月第一版　2019 年 6 月第二版　2025 年 2 月第三版
印　　次：2025 年 2 月北京第一次印刷
开　　本：787 毫米×1092 毫米　16 开本
印　　张：14.25
字　　数：340 千字
定　　价：46.60 元

版 权 专 有　侵 权 必 究

本书如有印装质量问题，我社营销中心负责退换

PREFACE
前言

随着新技术革命的发展及经济全球化的到来，社会对专业人才的外语能力要求越来越高。理工科的学生除了应具有一定的听说读写能力外，还应掌握一定的本专业基础词汇，具有基本的阅读本专业外文资料和进行专业交流的能力。为了更好地培养学生的专业英语应用能力，编者编写了本书。

为学习贯彻落实党的二十大精神，本书根据《党的二十大报告学习辅导百问》《二十大党章修正案学习问答》，在数字资源中设置了"二十大报告及党章修正案学习辅导"栏目，以方便师生学习。

在本次改版中，为了更好地适应专业建设需求，本书删减了部分内容，增加了供电技术章节。

本书由电工电子技术基础、电机与电机控制、工业计算机控制技术、供电技术和机器人技术5章组成，共计25课。电工电子技术基础主要介绍电路元器件、电路分析、二极管、运算放大器、逻辑电路等；电机与电机控制主要介绍直流电动机、交流异步电动机、电动机控制电路、变频器等；工业计算机控制技术主要介绍PLC、单片机、现场总线等；供电技术主要介绍发电、输电、配电等；机器人技术主要介绍机器人类型、机器人控制方法、嵌入式系统、机器人视觉系统等。

本书所有课文均精心选自国外相关网站和教材，在选材过程中，注意选择能体现我国技术优势的文章，以激发学生的民族自豪感；注意选择体现科学技术的文章，培养学生创新意识，以激励学生的文化自信。教材内容具有专业性和实用性强、难度适宜等特点。为了便于学生自学，所有课文均附有参考译文。

本书由王琳、汤慧芹主编，夏怡、吴梅任副主编，徐志成参加编写。本书承蒙南京师范大学陈雪丽教授审阅，提出了宝贵的修改意见，在此表示衷心感谢。

由于编者水平有限，书中难免存在疏漏之处，敬请读者批评指正。

编者
2024年11月

CONTENTS
目录

PREFACE

Chapter 1　Electrical and Electronic Technology Fundamentals ⋯⋯⋯⋯⋯1

　Unit 1　Electrical Elements and Network ⋯⋯⋯⋯⋯⋯⋯⋯⋯⋯⋯⋯⋯⋯ 2
　Reading 1　Circuit Diagram ⋯⋯⋯⋯⋯⋯⋯⋯⋯⋯⋯⋯⋯⋯⋯⋯⋯⋯⋯⋯ 6
　Unit 2　Network Analysis ⋯⋯⋯⋯⋯⋯⋯⋯⋯⋯⋯⋯⋯⋯⋯⋯⋯⋯⋯⋯⋯ 9
　Reading 2　Kirchhoff's Circuit Laws ⋯⋯⋯⋯⋯⋯⋯⋯⋯⋯⋯⋯⋯⋯⋯⋯ 13
　Unit 3　Diodes ⋯⋯⋯⋯⋯⋯⋯⋯⋯⋯⋯⋯⋯⋯⋯⋯⋯⋯⋯⋯⋯⋯⋯⋯⋯ 17
　Reading 3　Transistors ⋯⋯⋯⋯⋯⋯⋯⋯⋯⋯⋯⋯⋯⋯⋯⋯⋯⋯⋯⋯⋯ 24
　Unit 4　Operational Amplifier ⋯⋯⋯⋯⋯⋯⋯⋯⋯⋯⋯⋯⋯⋯⋯⋯⋯⋯ 28
　Reading 4　The Basics of Integrated Circuits ⋯⋯⋯⋯⋯⋯⋯⋯⋯⋯⋯⋯ 33
　Unit 5　Digital Logic Gates ⋯⋯⋯⋯⋯⋯⋯⋯⋯⋯⋯⋯⋯⋯⋯⋯⋯⋯⋯ 35
　Reading 5　Binary Numbers ⋯⋯⋯⋯⋯⋯⋯⋯⋯⋯⋯⋯⋯⋯⋯⋯⋯⋯⋯ 39
　Unit 6　Flip-flop ⋯⋯⋯⋯⋯⋯⋯⋯⋯⋯⋯⋯⋯⋯⋯⋯⋯⋯⋯⋯⋯⋯⋯⋯ 41
　Reading 6　Digital Counters ⋯⋯⋯⋯⋯⋯⋯⋯⋯⋯⋯⋯⋯⋯⋯⋯⋯⋯⋯ 45

Chapter 2　Electric Machines and Motor Control ⋯⋯⋯⋯⋯⋯⋯⋯⋯⋯ 48

　Unit 7　Motor and DC Motor ⋯⋯⋯⋯⋯⋯⋯⋯⋯⋯⋯⋯⋯⋯⋯⋯⋯⋯ 49
　Reading 7　Brushless DC Motor ⋯⋯⋯⋯⋯⋯⋯⋯⋯⋯⋯⋯⋯⋯⋯⋯⋯ 54
　Unit 8　Induction Motors ⋯⋯⋯⋯⋯⋯⋯⋯⋯⋯⋯⋯⋯⋯⋯⋯⋯⋯⋯⋯ 56
　Reading 8　Synchronous Machines: Generalities ⋯⋯⋯⋯⋯⋯⋯⋯⋯⋯ 62
　Unit 9　Simple Motor Control Circuits ⋯⋯⋯⋯⋯⋯⋯⋯⋯⋯⋯⋯⋯⋯ 65
　Reading 9　How to Wire a Motor Starter ⋯⋯⋯⋯⋯⋯⋯⋯⋯⋯⋯⋯⋯ 73
　Unit 10　Pulse Width Modulation Motor Control Theory ⋯⋯⋯⋯⋯⋯ 77
　Reading 10　Speed and Position Control of DC Motors ⋯⋯⋯⋯⋯⋯⋯ 83
　Unit 11　Variable Frequency Drive ⋯⋯⋯⋯⋯⋯⋯⋯⋯⋯⋯⋯⋯⋯⋯⋯ 86

Reading 11　Vector Control (Motor) ········· 91

Chapter 3　Industry Computer Control Technology ········· 95

Unit 12　Overview of Programmable Logic Controller (PLC) ········· 96
Reading 12　Process Control System with PLC ········· 101
Unit 13　Applications of PLC ········· 105
Reading 13　Ladder Diagram ········· 112
Unit 14　Introduction to Microcontrollers ········· 115
Reading 14　The Beginning ········· 121
Unit 15　8051 Instruction Set ········· 124
Reading 15　Basic I/O Operations of 8051 ········· 129
Unit 16　Fieldbus ········· 134
Reading 16　Profibus ········· 139

Chapter 4　Power Supply Technology ········· 144

Unit 17　Thermal Power Station ········· 145
Reading 17　Components of Thermal Power Plant ········· 148
Unit 18　Solar Energy ········· 152
Reading 18　What is Wind Energy ········· 159
Unit 19　Transmission and Distribution of Electricity ········· 164
Reading 19　What is Power Factor ········· 168
Unit 20　Difference between AC and DC Transmission System ········· 171
Reading 20　What are the Advantages of HVDC over HVAC ········· 174
Unit 21　Smart Grid ········· 177
Reading 21　Integration of Renewable Energy with Grid System ········· 181

Chapter 5　Robot Technology ········· 184

Unit 22　Types of Robot ········· 185
Reading 22　Arm Geometries ········· 190
Unit 23　Robot Control Methods ········· 192
Reading 23　Robot Control ········· 196
Unit 24　Embedded Systems for Robot ········· 199
Reading 24　Gear ········· 203

Unit 25　Robot Vision ···206
Reading 25　Sensors for Intelligent Robot ··························210
附录 1　电气专业课程词汇中英文对照 ·································213
附录 2　论文英文摘要的书写 ··215
参考文献 ···217

Chapter 1
Electrical and Electronic Technology Fundamentals

Unit 1　Electrical Elements and Network

The concept of electrical elements is used in the analysis of electrical networks. Any electrical network can be modeled by decomposing it down to multiple, interconnected electrical elements in a schematic diagram or circuit diagram[1]. Each electrical element affects the voltage in the network or current through the network in a particular way. By analyzing the way a network is affected by its individual elements, it is possible to estimate how a real network will behave on a macro scale.

Elements vs. components

There is a distinction between real, physical electrical or electronic components and the ideal electrical elements by which they are represented.
- Electrical elements do not exist physically, and are assumed to have ideal properties according to a lumped element model.
- Conversely, components do exist, have less than ideal properties, their values always have a degree of uncertainty, they always include some degree of nonlinearity and typically require a combination of multiple electrical elements to approximate their functions.

Circuit analysis using electrical elements is useful for understanding many practical electrical networks using components.

Elements

The four fundamental circuit variables are current, I; voltage, U; charge, Q; and magnetic flux, Φ_m. Only 5 elements are required to represent any component or network by manipulating these four variables.

Two sources:
- Current source, measured in amperes, produces a current in a conductor. Affects charge according to the relation $dQ = - I dt$.
- Voltage source, measured in volts, produces a potential difference between two points. Affects magnetic flux according to the relation $d\Phi_m = U dt$.

Three passive elements:
- Resistance R, measured in ohms, produces a voltage proportional to the current flowing through it. Relates voltage and current according to the relation $dU = RdI$.
- Capacitance C, measured in farads, produces a current proportional to the rate of change of voltage across it. Relates charge and voltage according to the relation $dQ = CdU$.
- Inductance L, measured in henries, produces a voltage proportional to the rate of change of current through it. Relates flux and current according to the relation $d\Phi_m = LdI$.

Examples

The following are examples of representation of components by way of electrical elements.

- On a first degree of approximation, a battery is represented by a voltage source. A more refined model also includes a resistance in series with the voltage source, to represent the battery's internal resistance (which results in the battery heating and the voltage dropping when in use). A current source in parallel may be added to represent its leakage (which discharges the battery over a long period of time).
- On a first degree of approximation, a resistor is represented by a resistance. A more refined model also includes a series inductance, to represent the effects of its lead inductance[2] (resistors constructed as a spiral have more significant inductance). A capacitance in parallel may be added to represent the capacitive effect of the proximity of the resistor leads to each other. A wire can be represented as a low-value resistor.
- Current sources are more often used when representing semiconductors. For example, on a first degree of approximation, a bipolar transistor may be represented by a variable current source that is controlled by the input voltage.

Electrical network

An electrical network is an interconnection of electrical elements such as resistors, inductors, capacitors, transmission lines, voltage sources, current sources, and switches.

An electrical circuit is a network that has a closed loop, giving a return path for the current (Fig.1.1). A network is a connection of two or more components, and may not necessarily be a circuit.

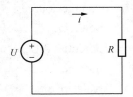

Fig.1.1 A simple electric circuit

Electrical networks that consist only of sources (voltage or current), linear lumped elements (resistors, capacitors, inductors), and linear distributed elements (transmission lines) can be analyzed by algebraic and transform methods to determine DC response, AC response, and transient response.

A network that also contains active electronic components is known as an electronic circuit. Such networks are generally nonlinear and require more complex design and analysis tools.

Technical Words and Expressions

electronic	[ilek'trɔnik]	adj.	电子的
electrical	[i'lektrik(ə)l]	adj.	电的，有关电的
element	['elimənt]	n.	元件
network	['netwə:k]	n.	网络，网状物
model	['mɔdl]	n.&v.	样式，模型；作…的模型
decompose	[,di:kəm'pəuz]	vt.	分解

multiple	[ˈmʌltipl]	adj.	多样的，多重的
schematic	[skiˈmætik]	adj.	示意性的
diagram	[ˈdaiəgræm]	n.	图样，图表，图像，示图
circuit	[ˈsəːkit]	n.	电路
schematic diagram			示意图，原理图
circuit diagram			电路图
macro	[ˈmækrəu]	adj.	巨大的，大量使用的
component	[kəmˈpəunənt]	n.	元件，组（部）件
lumped element model			集总元件模型
nonlinearity	[ˌnɔnliniˈæriti]	n.	非线性
approximate	[ˈprɔksimeit]	adj. & v.	近似的，大约的；近似，接近
variable	[ˈvɛəriəbl]	n.	变量
current	[ˈkʌrənt]	n.	电流
voltage	[ˈvəultidʒ]	n.	[电] 电压，伏特数
charge	[tʃɑːdʒ]	n. & v.	电荷；充电
magnetic	[mægˈnetik]	adj.	磁的，有磁性的，有吸引力的
flux	[flʌks]	n.	[物] 流量，通量
current source			电流源
ampere	[ˈæmpeə(r)]	n.	安培
conductor	[kənˈdʌktə]	n.	[物] 导体
voltage source			电压源
potential	[pəˈtenʃ(ə)l]	adj. & n.	势的，位的；电位
potential difference			电位差
passive	[ˈpæsiv]	adj.	无源的
resistance	[riˈzistəns]	n.	电阻，阻抗
ohm	[əum]	n.	[物] 欧姆
proportional	[prəˈpɔːʃnl]	adj.	（成）适当比例的，（与...）相称的（to）
capacitance	[kəˈpæsitəns]	n.	容量，电容
farad	[ˈfærəd]	n.	[电] 法拉（电容单位）
across	[əˈkrɔs]	prep.	跨接
inductance	[inˈdʌktəns]	n.	电感
henry	[ˈhenri]	n.	亨利（电感单位）
representation	[ˌreprizenˈteiʃən]	n.	表示法
approximation	[əˌprɔksiˈmeiʃən]	n.	[数] 近似值，近似法
battery	[ˈbætəri]	n.	电池
series	[ˈsiəriːz]	n.	[电] 串联（接）
parallel	[ˈpærəlel]	n.	[电] 并联
leakage	[ˈliːkidʒ]	n.	漏，泄漏，渗漏

Chapter 1 Electrical and Electronic Technology Fundamentals

discharge	[dis'tʃɑ:dʒ]	v.	[物] 放电
lead	[li:d]	n.	导线
spiral	['spaiərəl]	n.	螺线
capacitive	[kə'pæsitiv]	adj.	电容的
proximity	[prɔk'simiti]	n.	近似
semiconductor	['semikən'dʌktə]	n.	[物] 半导体
bipolar	[bai'pəulə]	adj.	有两极的，双极的
transistor	[træn'zistə]	n.	[电] 晶体管
switch	[switʃ]	n.	开关
loop	[lu:p]	n.	回路
distributed	[dis'tribju:tid]	adj.	分布式的
algebraic	[ˌældʒi'breiik]	adj.	代数的
transform method			变换法
response	[ris'pɔns]	n.	响应
transient	['trænziənt]	adj.	瞬时的
active	['æktiv]	adj.	有源的

💡 Notes

[1] Any electrical network can be modeled by decomposing it down to multiple, interconnected electrical elements in a schematic diagram or circuit diagram.

译文：任何电网络都可通过分解为原理图或电路图中多个相互连接的电气元件，来建立模型。

注释：此句中 any electrical network 译为任何电网络。

[2] A more refined model also includes a series inductance, to represent the effects of its lead inductance.

译文：更精确的模型中含有串联电感，可用来表示电阻导线中的电感效应。

注释：句中 to represent... 为目的状语。

 Exercises

I. **Mark the following statements with T (true) or F (false) according to the text.**

1. Each electrical element affects the voltage in the network or current through the network in a particular way. ( )

2. There isn't a distinction between real, physical electrical or electronic components and the ideal electrical elements by which they are represented. ( )

3. Components do exist, have some ideal properties. ( )
4. Voltage source, measured in volts, produces a potential difference between two points. ( )
5. Current sources are more often used when representing conductors. ( )

Ⅱ. **Complete the following sentences.**

1. The concept of _____ is used in the analysis of _____.
2. Each electrical element affects the _____ in the network or _____ through the network in a particular way.
3. Resistance R, measured in _____ produces a voltage _____ the current flowing through it.
4. on a first degree of approximation, a bipolar transistor may be represented by a _____ that is controlled by the _____.
5. A network is a connection of _____ or _____ components, and may not necessarily be a _____.

Reading 1 Circuit Diagram

A circuit diagram (also known as an electrical diagram, wiring diagram, elementary diagram, or electronic schematic) is a simplified conventional pictorial representation of an electrical circuit. It shows the components of the circuit as simplified standard symbols, and the power and signal connections between the devices. Arrangement of the components interconnections on the diagram does not correspond to their physical locations in the finished device.

Unlike a block diagram or layout diagram, a circuit diagram shows the actual wire connections being used. The diagram does not show the physical arrangement of components. A drawing meant to depict what the physical arrangement of the wires and the components they connect is called "artwork" or "layout" or the "physical design".

Circuit diagrams are used for the design (circuit design), construction (such as PCB layout), and maintenance of electrical and electronic equipment.

Legends

On a circuit diagram, the symbols for components are labelled with a descriptor (or reference designator) matching that on the list of parts. For example, C1 is the first capacitor, L1 is the first inductor, Q1 is the first transistor, and R1 is the first resistor (note that it isn't written R1, L1,...). The letters that precede the numbers were chosen in the early days of the electrical industry, even before the vacuum tube (thermionic valve), so "Q" was the only one available for semiconductor devices in the mid-twentieth century. Often the value or type designation of the component is given on the diagram beside the part, but detailed specifications would go on the parts list.

Chapter 1 Electrical and Electronic Technology Fundamentals

Symbols

Circuit diagram symbols (Fig.1.2) have differed from country to country and have changed over time, but are now to a large extent internationally standardized. Simple components often had symbols intended to represent some feature of the physical construction of the device. For example, the symbol for a resistor shown here dates back to the days when that component was made from a long piece of wire wrapped in such a manner as to not produce inductance, which would have made it a coil. These wire-wound resistors are now used only in high-power applications, smaller resistors being cast from carbon composition (a mixture of carbon and filler) or fabricated as an insulating tube or chip coated with a metal film. The internationally standardized symbol for a resistor is therefore now simplified to an oblong, sometimes with the value in ohms written inside, instead of the zigzag symbol. A less common symbol is simply a series of peaks on one side of the line representing the conductor, rather than back-and-forth as shown here.

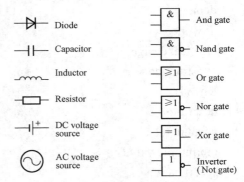

Fig.1.2 Circuit diagram symbols

Linkages

The linkages between leads were once simple crossings of lines; one wire insulated from and "jumping over" another was indicated by it making a little semicircle over the other line. With the arrival of computerized drafting, a connection of two intersecting wires was shown by a crossing with a dot or "blob", and a crossover of insulated wires by a simple crossing without a dot. However, there was a danger of confusing these two representations if the dot was drawn too small or omitted. Modern practice is to avoid using the "crossover with dot" symbol, and to draw the wires meeting at two points instead of one. It is also common to use a hybrid style, showing connections as a cross with a dot while insulated crossings use the semicircle. Several styles of schematic wire junctions are shown in Fig.1.3.

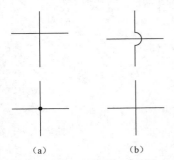

Fig.1.3 Schematic wire junctions
(a) Connection; (b) No connection

Technical Words and Expressions

wiring diagram		接线图
elementary diagram		原理图
pictorial	adj.	图示的

power	n.	电源
arrangement	n.	排列，布局
layout diagram		布置图
depict	vt.	描述，描写
artwork		布线图
physical design		布图设计
construction	n.	建造
legend	n.	图例
descriptor	n.	描述符
vacuum tube		真空管
thermionic valve		热离子管，热阴极电子管
specification	n.	规格，说明书
wrap	v.	卷，缠绕
cast	v.	浇铸
carbon composition		碳组分
filler	n.	填充剂
fabricate	v.	制造
insulating tube		绝缘管
metal film		金属膜
oblong	n.	长方形
zigzag	n.	锯齿形
peak	n.	波峰
back-and-forth		来回地，往复地
linkage	n.	连接
crossing	n.	交叉，相交
draft	vt.	设计
blob	n.	圆点
crossover	n.	交叉
hybrid	adj.	混合的
semicircle	n.	半圆形
junction	n.	连接，接合，交叉点，汇合处

Comprehension

1. A circuit diagram is also known as an electrical diagram, _____.
   A. wiring diagram   B. elementary diagram
   C. electronic schematic   D. all above

2. Unlike a block diagram or layout diagram, a circuit diagram shows the _____ wire connections being used.

A. virtual   B. actual   C. approximate   D. physical

3. According to the text, the distinction between a block diagram and a circuit diagram is that _____.

    A. a block diagram can show the physical arrangement of components
    B. a circuit diagram can show the physical arrangement of components
    C. a circuit diagram shows the actual wire connections being used
    D. a block diagram shows the actual wire connections being used

4. According to the text, which of the following description about circuit diagram is right? _____.

    A. Arrangement of the components interconnections on the diagram does correspond to their physical locations in the finished device
    B. There is no distinction between a circuit diagram and a layout diagram
    C. A circuit diagram can show the physical arrangement of components
    D. Circuit diagrams are used for the design (circuit design), construction (such as PCB layout), and maintenance of electrical and electronic equipments

5. The internationally standardized symbol for a resistor is therefore now simplified to _____, sometimes with the value in ohms written inside.

    A. a semicircle   B. an oblong   C. a triangle   D. a circle

Unit 2   Network Analysis

A network, in the context of electronics, is a collection of interconnected components. Network analysis is the process of finding the voltages across, and the currents through, every component in the network. There are a number of different techniques for achieving this. However, for the most part, they assume that the components of the network are all linear. The methods described in this article are only applicable to linear network analysis except where explicitly stated.

Definitions

Component: A device with two or more terminals into which, or out of which, charge may flow.

Node: A point at which terminals of more than two components are joined. A conductor with a substantially zero resistance is considered to be a node for the purpose of analysis.

Branch: The component (s) joining two nodes.

Mesh: A group of branches within a network joined so as to form a complete loop.

Port: Two terminals where the current into one is identical to the current out of the other.

Equivalent circuits

A useful procedure in network analysis is to simplify the network by reducing the number of components. This can be done by replacing the actual components with other notional components that have the same effect. A particular technique might directly reduce the number of components, for instance by combining impedances in series. On the other hand it might merely change the form into one in which the components can be reduced in a later operation. For instance, one might transform a voltage generator into a current generator using Norton's theorem in order to be able to later combine the internal resistance of the generator with a parallel impedance load.

A resistive circuit is a circuit containing only resistors, ideal current sources, and ideal voltage sources. If the sources are constant (DC) sources, the result is a DC circuit. The analysis of a circuit refers to the process of solving for the voltages and currents present in the circuit. The solution principles outlined here also apply to phasor analysis of AC circuits.

Two circuits (see circuit 1 and circuit 2 in Fig.1.4) are said to be equivalent with respect to a pair of terminals if the voltage across the terminals and current through the terminals for one network have the same relationship as the voltage and current at the terminals of the other network[1].

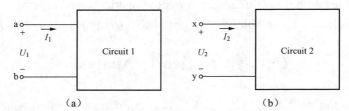

Fig.1.4　Circuit equivalence

If $U_2 = U_1$ implies $I_2 = I_1$ for all (real) values of U_1, then with respect to terminals ab and xy, circuit 1 and circuit 2 are equivalent.

The above is a sufficient definition for a one-port network. For more than one port, then it must be defined that the currents and voltages between all pairs of corresponding ports must bear the same relationship[2]. For instance, star and delta networks are effectively three port networks and hence require three simultaneous equations to fully specify their equivalence.

Impedances in series and in parallel

Any two terminal network of impedances can eventually be reduced to a single impedance by successive applications of impedances in series or impedances in parallel.

Impedances in series　　　　　　　$Z_{eq} = Z_1 + Z_2 + \cdots + Z_n$

Impedances in parallel　　　　　　　$\dfrac{1}{Z_{eq}} = \dfrac{1}{Z_1} + \dfrac{1}{Z_2} + \cdots + \dfrac{1}{Z_n}$

Chapter 1　Electrical and Electronic Technology Fundamentals

Source transformation

A generator with an internal impedance (ie non-ideal generator) can be represented as either an ideal voltage generator or an ideal current generator plus the impedance. These two forms (Fig.1.5) are equivalent and the transformations are given below. If the two networks are equivalent with respect to terminals ab, then U and I must be identical for both networks. Thus,

$$U_s = RI_s \quad \text{or} \quad I_s = \frac{U_s}{R}$$

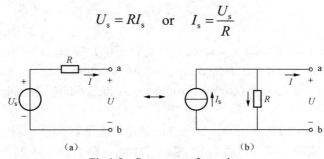

Fig.1.5　Source transformation
(a) Voltage source; (b) Current source

Technical Words and Expressions

in the context of			在…情况下
explicitly		adv.	明确地，明显地
terminal	[ˈtəːminl]	n.	终端，接线端
node	[nəud]	n.	节点
substantially	[səbˈstænʃ(ə)li]	adv.	实质上
branch	[brɑːntʃ]	n.	支路
mesh	[meʃ]	n.	网孔
port	[pɔːt]	n.	端口
notional	[ˈnəuʃənəl]	adj.	概念的，想象的
impedance	[imˈpiːdəns]	n.	阻抗
in series			串联
generator	[ˈdʒenəreitə]	n.	发电机，发生器
Norton's theorem			诺顿定理
load	[ləud]	n.	负载，负荷
resistive	[riˈzistiv]	adj.	电阻的，电阻性的
DC		n.	直流电
AC		n.	交流电
outline	[ˈəutlain]	v.	概述，略述
phasor	[ˈfeizə]	n.	相量
equivalent	[iˈkwivələnt]	adj.	等效的

one-port network			一端口网络
simultaneous	[ˌsiməl'teinjəs]	adj.	同时的，同时发生的
successive	[sək'sesiv]	adj.	连续的
in parallel			并联
transformation	[ˌtrænsfə'meiʃən]	n.	变换

Notes

[1] Two circuits (see circuit 1 and circuit 2 in Fig.1.4) are said to be equivalent with respect to a pair of terminals if the voltage across the terminals and current through the terminals for one network have the same relationship as the voltage and current at the terminals of the other network.

译文：如果一个网络的端电压和流过端点间的电流之间的关系与另一个网络端点间的电压和电流关系相同，这两个端点间的电路（图1.4中的电路1和电路2）等效。

注释：此句由if引导条件状语从句；主句为被动语态，在翻译的时候，可译成主动。

[2] For more than one port, then it must be defined that the currents and voltages between all pairs of corresponding ports must bear the same relationship.

译文：对于一端口以上的网络，应该定义所有相应的端口间具有相同的电流、电压关系。

注释：此句中it为形式主语，真正的主语为that所引导的从句。

Exercises

I. Mark the following statements with T (true) or F (false) according to the text.

1. A node is a point at which terminals of less than two components are joined.　　(　　)

2. A useful procedure in network analysis is to simplify the network by reducing the number of components.　　(　　)

3. The solution principles used for DC circuits can't apply to phasor analysis of AC circuits.　　(　　)

4. Any two terminal network of impedances can eventually be reduced to a single impedance by successive applications of impedances in series or impedances in parallel.　　(　　)

5. A generator with an internal impedance is an ideal generator.　　(　　)

II. Complete the following sentences.

1. Network analysis is the process of finding the _____ across, and the _____ through, every _____ in the network.

2. A useful procedure in network analysis is to _____ the network by reducing the _____ of components.

3. A resistive circuit is a circuit containing only _____, ideal _____, and ideal _____.

4. Star and delta networks are effectively _____ networks and hence require _____ simultaneous equations to fully specify their_____.

5. A generator with an _____ (ie non-ideal generator) can be represented as either _____ or _____ plus the impedance.

Reading 2 Kirchhoff's Circuit Laws

Kirchhoff's circuit laws are two equalities that deal with the conservation of charge and energy in electrical circuits, and were first described in 1845 by Gustav Kirchhoff. Widely used in electrical engineering, they are also called Kirchhoff's rules or simply Kirchhoff's laws (see also Kirchhoff's laws for other meanings of that term).

Both circuit rules can be directly derived from Maxwell's equations, but Kirchhoff preceded Maxwell and instead generalized work by Georg Ohm.

Kirchhoff's Current Law (KCL)

This law is also called Kirchhoff's first law, Kirchhoff's point rule, Kirchhoff's junction rule (or nodal rule), and Kirchhoff's first rule.

The principle of conservation of electric charge implies that:

At any point in an electrical circuit that does not represent a capacitor plate, the sum of currents flowing towards that point is equal to the sum of currents flowing away from that point.

Adopting the convention that every current flowing towards the point is positive and that every current flowing away is negative (or the other way around), this principle can be stated as

$$\sum_{k=1}^{n} I_k = 0$$

n is the total number of currents flowing towards or away from the point. This formula is also valid for complex currents

$$\sum_{k=1}^{n} \tilde{I}_k = 0$$

Applying Kirchhoff's Current Law to the diagram shown below (Fig.1.6), the current entering any junction is equal to the current leaving that junction: $i_1 + i_4 = i_2 + i_3$.

Changing charge density

Physically speaking, the restriction regarding the "capacitor plate" means that Kirchhoff's Current Law is only valid if the charge density remains constant in the point that it is applied to. This is

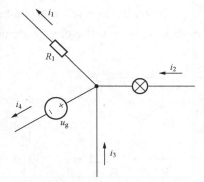

Fig.1.6 A diagram of Kirchhoff's Current Law

normally not a problem because of the strength of electrostatic forces: the charge buildup would cause repulsive forces to disperse the charges.

However, a charge build-up can occur in a capacitor, where the charge is typically spread over wide parallel plates, with a physical break in the circuit that prevents the positive and negative charge accumulations over the two plates from coming together and cancelling. In this case, the sum of the currents flowing into one plate of the capacitor is not zero, but rather is equal to the rate of charge accumulation. However, if the displacement current dD/dt is included, Kirchhoff's Current Law once again holds. (This is really only required if one wants to apply the current law to a point on a capacitor plate. In circuit analyses, however, the capacitor as a whole is typically treated as a unit, in which case the ordinary current law holds since exactly the current that enters the capacitor on the one side leaves it on the other side.)

Kirchhoff's Voltage Law (KVL)

This law is also called Kirchhoff's second law, Kirchhoff's loop (or mesh) rule, and Kirchhoff's second rule.

The directed sum of the electrical potential differences around any closed circuit must be zero.

Similarly to KCL, it can be stated as

$$\sum_{k=1}^{n} U_k = 0$$

Here, n is the total number of voltages measured. The voltages may also be complex

$$\sum_{k=1}^{n} \widetilde{U}_k = 0$$

Direction of the voltages

This law often leads to confusion due to sign errors.

Typically, in a schematic diagram, one has to choose whether to define a voltage measured clockwise or counter-clockwise as positive. It is a question of convenience, but once decided, all voltages must be treated this way.

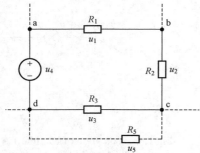

Fig.1.7 A diagram depicting Kirchhoff's Voltage Law

The confusion comes from the direction of the voltage at voltage sources and capacitors.

Considering the schematic diagram (Fig.1.7), we may decide to measure all voltages clockwise, that is, from a to b, from b to c, from c to d and from d to a. The sum of all the voltages around the loop is equal to zero, $u_1+u_2+u_3+u_4=0$.

At the resistors, the voltage will be positive, because we measure from "plus to minus". But when we measure from d to a, we will measure a negative voltage, because we measure from "minus to plus". We have to do this because otherwise, we would be measuring counter-clockwise if we measured the other way around.

If our voltage source is a capacitor, we will also notice that we must consider the charge negative, which is due to: $Q = CU$.

There is the rule of thumb that one must just "reverse supply voltages" in order to make the calculation match KVL. There is, however, the risk that this may lead to additional confusion in more complex circuits, in which it is not clear what the supply voltage exactly is.

It is actually not a contradiction that we measure from "minus to plus"; inside a voltage source, this is the direction in which the current flows. If a magnetic field induces a voltage into an inductor, making it a voltage source (which is the principle of an electrical generator), this should be obvious. Even in a battery, the circuit is closed because of Ions flowing and representing charge carriers.

Inside capacitors, there is no flow of charge carriers. The KVL is valid anyway, but we have to consider the electric field inside the capacitor in order to see this.

Electric field and electric potential

Kirchhoff's Voltage Law as stated above is equivalent to the statement that a single-valued electric potential can be assigned to each point in the circuit (in the same way that any conservative vector field can be represented as the gradient of a scalar potential).

This could be viewed as a consequence of the principle of conservation of energy. Otherwise, it would be possible to build a perpetual motion machine that passed a current in a circle around the circuit.

Considering that electric potential is defined as a line integral over an electric field, Kirchhoff's Voltage Law can be expressed equivalently as

$$\oint_C E dl = 0$$

which states that the line integral of the electric field around closed loop C is zero.

In order to return to the more special form, this integral can be "cut in pieces" in order to get the voltage at specific components.

This is a simplification of Faraday's law of induction for the special case where there is no fluctuating magnetic field linking the closed loop. Therefore, it practically suffices for explaining circuits containing only resistors and capacitors.

In the presence of a changing magnetic field the electric field is not conservative and it cannot therefore define a pure scalar potential —— the line integral of the electric field around the circuit is not zero. This is because energy is being transferred from the magnetic field to the current (or vice versa). In order to "fix" Kirchhoff's Voltage Law for circuits containing inductors, an effective potential drop, or electromotive force (emf), is associated with each inductance of the circuit, exactly equal to the amount by which the line integral of the electric field is not zero by Faraday's law of induction.

Technical Words and Expressions

equality	n.	等式
conservation	n.	守恒
energy	n.	能量
derive	vt.	起源于，出自
generalize	vt.	归纳，概括，推广
capacitor plate		电容器极板
be equal to		等于
positive	adj.	正的
negative	adj.	负的
formula	n.	公式
complex	adj	[数]复数的
density	n.	[物]浓度，密度
restriction	n.	限制，约束
electrostatic force		静电力
repulsive	adj.	推斥的，排斥的
disperse	v.	（使）分散，（使）散开
build-up	n.	增强，增长；上升
accumulation	n.	积聚
cancel	vt.	取消
displacement current		位移电流
confusion	n.	混乱，混淆
clockwise	adj. & adv.	顺时针方向的；顺时针方向地
counter-clockwise	adj. & adv.	逆时针方向的；逆时针方向地
convenience	n.	便利，方便
thumb	n.	拇指
contradiction	n.	矛盾
ion	n.	离子
charge carrier		电荷载体，载荷子
vector field		矢（向）量场
gradient	n.	梯度，倾斜度，坡度
scalar potential		标量势、标量位
consequence	n.	结果
perpetual motion machine		永动机
line integral		线积分
suffice	vi.	足够，有能力
electromotive force (emf)		电动势

Chapter 1 Electrical and Electronic Technology Fundamentals

Comprehension

1. Kirchhoff's Circuit Laws are two equalities that deal with the conservation of _____ in electrical circuits.
    A. current and voltage
    B. current and energy
    C. charge and energy
    D. charge and voltage
2. Applying Kirchhoff's Current Law to the diagram shown in Fig.1.6, the relation of the four currents can be written as _____.
    A. $i_1 + i_2 = i_3 + i_4$
    B. $i_1 + i_4 = i_2 + i_3$
    C. $i_1 + i_3 = i_2 + i_4$
    D. $i_1 = i_2 + i_3 + i_4$
3. Kirchhoff's Voltage Law can be summarized as follows: _____.
    A. around any closed loop in a circuit, the sum of all changes in potential (emfs and potential drops across resistors and other circuit elements) must equal zero
    B. around any closed loop in a circuit, the sum of all changes in potential (emfs and potential drops across resistors and other circuit elements) mustn't equal zero
    C. at each instant of time, the algebraic sum of the voltage rise is more than the algebraic sum of the voltage drops, both being taken in the same direction around the closed loop
    D. at each instant of time, the algebraic sum of the voltage rise is less than the algebraic sum of the voltage drops, both being taken in the same direction around the closed loop
4. Kirchhoff's Voltage Law can be applied only to _____.
    A. nodes   B. junctions   C. branches   D. closed loops
5. At the resistors, the voltage will be _____, because we measure from "plus to minus".
    A. positive   B. negative   C. zero   D. uncertain

Unit 3 Diodes

In electronics, a diode is a two-terminal device. Diodes have two active electrodes between which the signal of interest may flow, and most are used for their unidirectional electric current property. The varicap diode is used as an electrically adjustable capacitor.

The directionality of current flow most diodes exhibit is sometimes generically called the rectifying property. The most common function of a diode is to allow an electric current to pass in one direction (called the forward biased condition) and to block the current in the opposite direction (the reverse biased condition). Thus, the diode can be thought of as an electronic version of a check valve.

Real diodes do not display such a perfect on-off directionality but have a more complex non-linear electrical characteristic, which depends on the particular type of diode technology.

Diodes also have many other functions in which they are not designed to operate in this on-off manner.

Early diodes included "cat's whisker" crystals and vacuum tube devices. Today the most common diodes are made from semiconductor materials such as silicon or germanium.

Semiconductor diodes

Most modern diodes are based on semiconductor P-N junctions. In a P-N diode, conventional current can flow from the P-type side (the anode) to the N-type side (the cathode), but cannot flow in the opposite direction. Another type of semiconductor diode, the Schottky diode, is formed from the contact between a metal and a semiconductor rather than by a P-N junction.

Current-voltage characteristic

A semiconductor diode's current-voltage characteristic, or i–u curve, is related to the transport of carriers through the so-called depletion layer or depletion region that exists at the P-N junction between differing semiconductors [1]. When a P-N junction is first created, conduction band (mobile) electrons from the N-doped region diffuse into the P-doped region where there is a large population of holes (places for electrons in which no electron is present) with which the electrons "recombine". When a mobile electron recombines with a hole, both hole and electron vanish, leaving behind an immobile positively charged donor on the N-side and negatively charged acceptor on the P-side. The region around the P-N junction becomes depleted of charge carriers and thus behaves as an insulator.

However, the depletion width cannot grow without limit. For each electron-hole pair that recombines, a positively-charged dopant ion is left behind in the N-doped region, and a negatively charged dopant ion is left behind in the P-doped region. As recombination proceeds and more ions are created, an increasing electric field develops through the depletion zone which acts to slow and then finally stop recombination. At this point, there is a "built-in" potential across the depletion zone.

If an external voltage is placed across the diode with the same polarity as the built-in potential, the depletion zone continues to act as an insulator, preventing any significant electric current flow. This is the reverse bias phenomenon. However, if the polarity of the external voltage opposes the built-in potential, recombination can once again proceed, resulting in substantial electric current through the P-N junction. For silicon diodes, the built-in potential is approximately 0.6V. Thus, if an external current is passed through the diode, about 0.6V will be developed across the diode such that the P-doped region is positive with respect to the N-doped region and the diode is said to be "turned on" as it has a forward bias.

A diode's i–u characteristic can be approximated by four regions of operation (Fig.1.8).

At very large reverse bias, beyond the peak inverse voltage or PIV, a process called reverse breakdown occurs which causes a large increase in current that usually damages the device permanently. The avalanche diode is deliberately designed for use in the avalanche region. In the zener diode, the concept of PIV is not applicable. A zener diode contains a heavily doped P-N junction allowing electrons to tunnel from the valence band of the P-type material to the conduction band of the N-type material, such that the reverse voltage is "clamped" to a known value (called the zener voltage), and avalanche does not occur. Both devices, however, do have a limit to the maximum current and power in the clamped reverse voltage region.

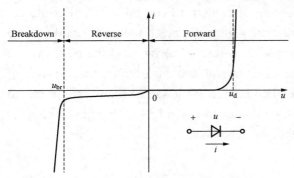

Fig.1.8　i–u characteristics of a P-N junction diode

The second region, at reverse biases more positive than the PIV, has only a very small reverse saturation current. In the reverse bias region for a normal P-N rectifier diode, the current through the device is very low (in the μA range).

The third region is forward but small bias, where only a small forward current is conducted.

As the potential difference is increased above an arbitrarily defined "cut-in voltage" or "on-voltage" or "diode forward voltage drop (U_d)", the diode current becomes appreciable (the level of current considered "appreciable" and the value of cut-in voltage depends on the application), and the diode presents a very low resistance.

The current-voltage curve is exponential. In a normal silicon diode at rated currents, the arbitrary "cut-in" voltage is defined as 0.6 to 0.7V. The value is different for other diode types —— Schottky diodes can be as low as 0.2V and red light-emitting diodes (LEDs) can be 1.4V or more and blue LEDs can be up to 4.0V.

At higher currents the forward voltage drop of the diode increases. A drop of 1 to 1.5V is typical at full rated current for power diodes.

Types of semiconductor diode

There are several types of junction diodes (Fig.1.9), which either emphasize a different physical aspect of a diode often by geometric scaling, doping level, choosing the right electrodes, are just an application of a diode in a special circuit, or are really different devices like the Gunn, laser diode and the MOSFET.

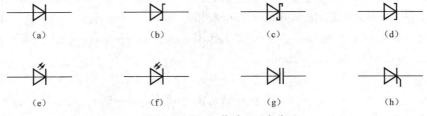

Fig.1.9　Some diode symbols

(a) Diode; (b) Zener diode; (c) Schottky diode; (d) Tunnel diode; (e) Light-emitting diode;
(f) Photodiode; (g) Varicap; (h) Silicon controlled rectifier

Normal (P-N) diodes, which operate as described above, are usually made of doped silicon or, more rarely, germanium. Before the development of modern silicon power rectifier diodes, cuprous and later selenium was used; its low efficiency gave it a much higher forward voltage drop (typically 1.4~1.7V per "cell", with multiple cells stacked to increase the peak inverse voltage rating in high voltage rectifiers), and required a large heat sink (often an extension of the diode's metal substrate), much larger than a silicon diode of the same current ratings would require. The vast majority of all diodes are the P-N diodes found in CMOS integrated circuits, which include two diodes per pin and many other internal diodes.

Applications

Radio demodulation

The first use for the diode is the demodulation of amplitude modulated (AM) radio broadcasts. The history of this discovery is treated in depth in the radio article. In summary, an AM signal consists of alternating positive and negative peaks of voltage, whose amplitude or "envelope" is proportional to the original audio signal. The diode (originally a crystal diode) rectifies the AM radio frequency signal, leaving an audio signal which is the original audio signal. The audio is extracted using a simple filter and fed into an audio amplifier or transducer, which generates sound waves.

Power conversion

Rectifiers are constructed from diodes, where they are used to convert alternating current (AC) electricity into direct current (DC). Automotive alternators are a common example, where the diode, which rectifies the AC into DC, provides better performance than the commutator of earlier dynamo [2]. Similarly, diodes are also used in Cockcroft-Walton voltage multipliers to convert AC into higher DC voltages.

Over-voltage protection

Diodes are frequently used to conduct damaging high voltages away from sensitive electronic devices. They are usually reverse-biased (nonconducting) under normal circumstances. When the voltage rises above the normal range, the diodes become forward-biased (conducting). For example, diodes are used in (stepper motor and H-bridge) motor controller and relay circuits to de-energize coils rapidly without the damaging voltage spikes that would otherwise occur[3]. (Any diode used in such an application is called a flyback diode.) Many integrated circuits also incorporate diodes on the connection pins to prevent external voltages from damaging their sensitive transistors. Specialized diodes are used to protect from over-voltages at higher power (see Diode types above).

Technical Words and Expressions

diode	['daiəud]	n.	二极管
electrode	[i'lektrəud]	n.	电极

Chapter 1 Electrical and Electronic Technology Fundamentals

unidirectional	[ˌjuːnɪdîˈrekʃənəl]	adj.	单向的，单向性的
varicap	[ˌværiˈkæp]	n.	变容二极管
varicap diode			变容二极管
directionality	[diˌrekʃənˈnæliti]	n.	方向性
rectify	[ˈrektifai]	vt.	[电] 把…整流
forward biased			正偏的
reverse biased			反偏的
check valve			单向阀
silicon	[ˈsilikən]	n.	[化] 硅，硅元素
germanium	[dʒəːˈmeiniəm]	n.	锗
P-N junction			PN 结
conventional	[kənˈvenʃənl]	adj.	惯例的，常规的
Schottky diode			肖特基二极管
current-voltage characteristic			伏安特性
depletion	[diˈpliːʃən]	n.	耗散
doped	[dəupt]	adj.	掺杂质的
N-doped region			N 型掺杂区
hole	[həul]	n.	[物] 空穴
electron	[iˈlektrɔn]	n.	电子
immobile	[iˈməubail]	adj.	静止的
donor	[ˈdəunə]	n.	供体
acceptor	[əkˈseptə(r)]	n.	受体
deplete	[diˈpliːt]	vt.	耗尽
dopant	[ˈdəupənt]	n.	掺杂物，掺杂剂
substantial	[səbˈstænʃəl]	adj.	实质的，真实的
breakdown	[ˈbreikdaun]	n.	[电] 击穿
avalanche	[ˈævəˌlɑːnʃ]	n.	雪崩
avalanche diode			雪崩二极管
deliberately		adv.	故意地
zener diode			稳压二极管
tunnel	[ˈtʌnl]	n.	隧道，地道
valence band			价电子带
tunnel diode			隧道二极管
clamp	[klæmp]	vt.	钳位
saturation	[ˌsætʃəˈreiʃən]	n.	饱和（状态）
cut-in voltage			开启电压；阈值电压
appreciable	[əˈpriːʃiəbl]	adj.	可感知的
exponential	[ˌekspəuˈnenʃəl]	adj.	指数的，幂数的
light-emitting diode (LED)			发光二极管

geometric scaling			几何尺寸
Gunn		adj.	[电] 耿氏效应的，基于耿氏效应的
cuprous oxide			氧化亚铜
selenium	[si'li:niəm]	n.	[化] 硒
heat sink			散热片，散热装置
extension	[iks'tenʃən]	n.	扩大，伸展
substrate	['sʌbstreit]	n.	衬底，基片，基体
integrated circuit			集成电路
pin	[pin]	n.	引脚，管脚
anode	['ænəud]	n.	[电] 阳极，正极
cathode	['kæθəud]	n.	阴极
demodulation	['di:,mɔdju:'leiʃən]	n.	[电] 解调，检波
amplitude modulated (AM) radio broadcast			调幅无线广播
envelope	['enviləup]	n.	包迹[线]，包络（线，面）
rectifier	['rektifaiə]	n.	整流器
audio	['ɔ:diəu]	adj.	音频的
frequency	['fri:kwənsi]	n.	频率
extract	[iks'trækt]	vt.	推断出，引出
filter	['filtə]	n.	滤波器
automotive alternator			机动车交流发电机
commutator	['kɔmjuteitə]	n.	换向器，转接器
dynamo	['dainəməu]	n.	发电机
Cockcroft-Walton voltage multiplier			科克罗夫-沃尔电压倍增器
stepper motor			步进电动机
H-bridge			H 桥电路
relay	['ri:lei]	n.	继电器
de-energize			去激励
spike	[spaik]	n.	尖峰信号，测试信号
flyback diode			回扫二极管

Notes

[1] A semiconductor diode's current-voltage characteristic, or $i-u$ curve, is related to the transport of carriers through the so-called depletion layer or depletion region that exists at the P-N junction between differing semiconductors.

译文：半导体二极管的电流-电压特性，即 $i-u$ 曲线，与载流子流过的耗散层或耗散区有关，耗散层或耗散区位于不同类型半导体间形成的 PN 结上。

注释：句中 that 引导定语从句修饰 depletion layer 和 depletion region。

Chapter 1 Electrical and Electronic Technology Fundamentals

[2] Automotive alternators are a common example, where the diode, which rectifies the AC into DC, provides better performance than the commutator of earlier dynamo.

译文：机动车交流发电机是一个常用的例子，其中二极管把交流变为直流，比早期的发电机换向器提供更好的性能。

注释：句中 where 引导非限制性定语从句，从句的谓语为 provides。

[3] For example, diodes are used in (stepper motor and H-bridge) motor controller and relay circuits to de-energize coils rapidly without the damaging voltage spikes that would otherwise occur.

译文：例如，二极管用在（步进电动机和 H 桥电路）电动机控制器和继电器电路中对线圈快速地去激励时，不会产生可能出现的具有危害性的电压尖峰。

注释：句中 to 引导目的状语。

Exercises

Ⅰ. **Mark the following statements with T (true) or F (false) according to the text.**

1. Diodes are used for their bidirectional electric current property. (　　)
2. Today the most common diodes are made from semiconductor materials such as silicon or germanium. (　　)
3. In a P-N diode, conventional current can flow from the N-type side (the cathode) to the p-type side (the anode). (　　)
4. Rectifiers are constructed from diodes, where they are used to convert alternating current (AC) electricity into direct current (DC). (　　)
5. Many integrated circuits also incorporate diodes on the connection pins to prevent external voltages from damaging their sensitive transistors. (　　)

Ⅱ. **Complete the following sentences.**

1. Most modern diodes are based on semiconductor _____. In a P-N diode, conventional current can flow from the _____ side (the anode) to the _____ side (the cathode), but cannot flow in the _____ direction.

2. If an external voltage is placed across the diode with the same polarity as the _____ potential, the depletion zone continues to act as an _____, preventing any significant electric current flow.

3. In a normal silicon diode at rated currents, the arbitrary "cut-in" voltage is defined as _____ to _____ Volts.

4. The audio is extracted using a simple _____ and fed into an _____ or transducer, which generates sound waves.

5. _____ diodes are used to protect from _____ at higher power.

Reading 3 Transistors

In electronics, a transistor is a semiconductor device commonly used to amplify or switch electronic signals. A transistor is made of a solid piece of a semiconductor material, with at least three terminals for connection to an external circuit. A voltage or current applied to one pair of the transistor's terminals changes the current flowing through another pair of terminals. Because the controlled (output) power can be much larger than the controlling (input) power, the transistor provides amplification of a signal. The transistor is the fundamental building block of modern electronic devices, and is used in radio, telephone, computer and other electronic systems. Some transistors are packaged individually but most are found in integrated circuits.

Types (Fig.1.10)

Transistors are categorized by:
- Semiconductor material : germanium, silicon, gallium arsenide, silicon carbide, etc.
- Structure: BJT, JFET, IGFET (MOSFET), IGBT, "other types".
- Polarity: NPN, PNP (BJTs); N-channel, P-channel (FETs).
- Maximum power rating: low, medium, high.
- Maximum operating frequency: low, medium, high, radio frequency (RF), microwave. (The maximum effective frequency of a transistor is denoted by the term f_T, an abbreviation for "frequency of transition". The frequency of transition is the frequency at which the transistor yields unity gain.)
- Application: switch, general purpose, audio, high voltage, super-beta, matched pair.
- Physical packaging: through hole metal, through hole plastic, surface mount, ball grid array, power modules.
- Amplification factor h_{fe} (transistor beta).

Thus, a particular transistor may be described as: silicon, surface mount, BJT, NPN, low power, high frequency switch.

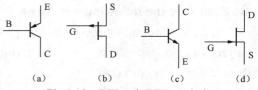

Fig.1.10 BJT and JFET symbols
(a) PNP; (b) P-channel; (c) NPN; (d) N-channel

Usage

In the early days of transistor circuit design, the bipolar junction transistor, or BJT, was the most commonly used transistor. Even after MOSFETs became available, the BJT remained the transistor of choice for digital and analog circuits because of their ease of manufacture and speed. However, desirable properties of MOSFETs, such as their utility in low-power devices, have made them the ubiquitous choice for use in digital circuits and a very common choice for use in analog circuits.

Transistor as a switch

Transistors are commonly used as electronic switches, for both high power applications including switched-mode power supplies and low power applications such as logic gates.

It can be seen from the graph (Fig.1.11) that once the base voltage reaches a certain level, shown at B, no more current will exist and the output will be held at a fixed voltage. The transistor is then said to be saturated. Hence, values of input voltage can be chosen such that the output is either completely off, or completely on. The transistor is acting as a switch, and this type of operation is common in digital circuits where only "on" and "off" values are relevant.

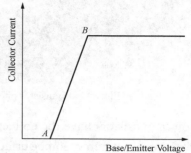

Fig.1.11　Operation graph of a transistor

Transistor as an amplifier

The below common emitter amplifier (Fig.1.12) is designed so that a small change in voltage in (U_{in}) changes the small current through the base of the transistor and the transistor's current amplification combined with the properties of the circuit mean that small swings in U_{in} produce large changes in U_{out}.

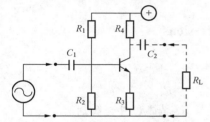

Fig.1.12　Common emitter amplifier

It is important that the operating parameters of the transistor are chosen and the circuit designed such that as far as possible the transistor operates within a linear portion of the graph, otherwise the output signal will suffer distortion.

Various configurations of single transistor amplifier are possible, with some providing current gain, some voltage gain, and some both.

From mobile phones to televisions, vast numbers of products include amplifiers for sound reproduction, radio transmission, and signal processing. The first discrete transistor audio amplifiers barely supplied a few hundred milliwatts, but power and audio fidelity gradually increased as better transistors became available and amplifier architecture evolved.

Modern transistor audio amplifiers of up to a few hundred watts are common and relatively inexpensive. Transistors have replaced valves (electron tubes) in instrument amplifiers.

Some musical instrument amplifier manufacturers mix transistors and vacuum tubes in the same circuit, as some believe tubes have a distinctive sound.

Packaging

Transistors come in many different packages (chip carriers) (Fig.1.13). The two main categories are through-hole (or leaded), and surface-mount, also known as surface mount device (SMD). The ball grid array (BGA) is the latest surface mount package (currently only for large transistor arrays). It has solder "balls" on the underside in place of leads. Because they are smaller and

Fig.1.13 Transistor photo

have shorter interconnections, SMDs have better high frequency characteristics but lower power rating.

Transistor packages are made of glass, metal, ceramic or plastic. The package often dictates the power rating and frequency characteristics. Power transistors have large packages that can be clamped to heat sinks for enhanced cooling. Additionally, most power transistors have the collector or drain physically connected to the metal can/metal plate. At the other extreme, some surface-mount microwave transistors are as small as grains of sand.

Often a given transistor type is available in different packages. Transistor packages are mainly standardized, but the assignment of a transistor's functions to the terminals is not: different transistor types can assign different functions to the package's terminals. Even for the same transistor type the terminal assignment can vary (normally indicated by a suffix letter to the part number, i.e. BC212L and BC212K).

Technical Words and Expressions

solid	adj.	固体（态）的
amplification	n.	放大
package	n. & vt.	晶体管外壳；封装
categorize	v.	把…分类，分类
gallium arsenide		砷化镓
carbide	n.	［化］碳化物
polarity	n.	极性
channel	n.	沟道
maximum power rating		最大额定功率
radio frequency		射频
abbreviation	n.	缩写，缩写词
matched pair		配对
through hole		通孔，穿孔
surface mount		表面安装
ball grid array		球栅阵列（封装）
power module		电源模块
bipolar	adj.	双极的
ubiquitous	adj.	普遍存在的
switched-mode power supply		开关模式电源
logic gate		逻辑门

Chapter 1 Electrical and Electronic Technology Fundamentals

base	n.	基极
emitter	n.	发射极
common emitter amplifier		共发射极放大器
parameter	n.	参数
gain	n.	增益
mobile phone		移动电话
discrete	adj.	不连续的，离散的
barely	adv.	仅仅，刚刚，几乎不能
milliwatt	n.	毫瓦［特］
fidelity	n.	保真度
evolve	v.	（使）发展，（使）进展
musical	adj.	音乐的，悦耳的
chip carrier		晶片载体，芯片载体
dictate	v.	规定
ceramic	adj.	陶瓷的，陶器的
collector	n.	集电极
drain	n.	漏极
grain	n.	晶粒
assignment	n.	排布
suffix	n.	后缀，下标

Comprehension

1. Transistors are commonly used as electronic _____, for both high power applications including switched-mode power supplies and low power applications such as logic gates.
    A. heaters   B. transformers   C. amplifiers   D. switches

2. This arrangement where the emitter (E) is in the controlling circuit (base current) and in the controlled circuit (collector current) is called _____.
    A. common collector mode   B. common base mode
    C. common emitter mode   D. all above

3. Which of the following statements is not true about functional model of transistor? _____.
    A. The base-emitter junction behaves like a diode
    B. The small base current I_B controls the large collector current I_C
    C. The collector current I_C is controlled by the base current I_B
    D. A base current I_B flows only when the voltage U_{BE} across the base-emitter junction is less than 0.7V

4. When a transistor works as an amplifier, its larger collector current I_C is _____ the base current I_B.
    A. more than   B. less than   C. proportional to   D. equal to

5. The transistor should operate within a _____ portion of the graph, otherwise the output signal will suffer distortion.

    A. nonlinear   B. linear   C. cut-off   D. saturation

Unit 4　Operational Amplifier

An operational amplifier, often called an op-amp, is a DC-coupled high-gain electronic voltage amplifier with differential inputs and, usually, a single output. Typically the output of the op-amp is controlled either by negative feedback, which largely determines the magnitude of its output voltage gain, or by positive feedback, which facilitates regenerative gain and oscillation[1]. High input impedance at the input terminals and low output impedance are important typical characteristics.

Op-amps are among the most widely used electronic devices today, being used in a vast array of consumer, industrial, and scientific devices. Many standard IC op-amps cost only a few cents in moderate production volume; however some integrated or hybrid operational amplifiers with special performance specifications may cost over $100 in small quantities.

Modern designs are electronically more rugged than earlier implementations and some can sustain direct short-circuits on their outputs without damage.

Circuit notation

The circuit symbol for an op-amp is shown in Fig.1.14 where:
- U_+: non-inverting input;
- U_-: inverting input;
- U_{out}: output;
- U_{s+}: positive power supply;
- U_{s-}: negative power supply.

The power supply pins (U_{s+} and U_{s-}) can be labeled in different ways. Despite different labeling, the function remains the same. Often these pins are left out of the diagram for clarity, and the power configuration is described or assumed from the circuit. The positions of the inverting and non-inverting inputs may be reversed in diagrams where appropriate; the power supply pins are not commonly reversed. For Example IC741 is an operational amplifier. It is used for doing arithmetic operations on analog computers, instrumentation and other control systems. Operational amplifier is in the class of linear ICs. Linear have a peculiarity that they can take continuous voltage signals like their analog counterparts. These are highly used today because of their high reliability and low cost. They are mainly used as voltage amplifiers. The basic operational amplifier works similar

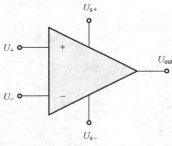

Fig.1.14　Op-amp symbol

to the following sequence:

input stage→intermediate stage→level shifter→output stage.

Input stage consists of high input impedance it amplifies the difference between the given input signals. The intermediate stage consists of cascaded amplifiers to amplify the signals from the input. Due to high amplification the DC level of the signals goes up. So in order to bring them down to the rated value, level shifter or level translator is used. The output stage consists of class AB/ class B power amplifier in order to amplify the power of the output signal.

Operation of ideal op-amps

The amplifier's differential inputs consist of an inverting input and a non-inverting input, and ideally the op-amp amplifies only the difference in voltage between the two. This is called the "differential input voltage". In its most common use, the op-amp's output voltage is controlled by feeding a fraction of the output signal back to the inverting input. This is known as negative feedback. If that fraction is zero, i.e., there is no negative feedback, the amplifier is said to be running "open loop" and its output is the differential input voltage multiplied by the total gain of the amplifier, as shown by the following equation

$$U_{out}=(U_+-U_-)G$$

where U_+ is the voltage at the non-inverting terminal, U_- is the voltage at the inverting terminal and G is the total open-loop gain of the amplifier.

Because the magnitude of the open-loop gain is typically very large and not well controlled by the manufacturing process, op-amps are not usually used without negative feedback. Unless the differential input voltage is extremely small, open-loop operation results in op-amp saturation. An example of how the output voltage is calculated when negative feedback exists is shown below in basic non-inverting amplifier circuit.

Another typical configuration of op-amps is the positive feedback, which takes a fraction of the output signal back to the non-inverting input. An important application of it is the comparator with hysteresis.

For any input voltages the ideal op-amp has:
- infinite open-loop gain;
- infinite bandwidth;
- infinite input impedances (resulting in zero input currents);
- zero offset voltage;
- infinite slew rate;
- zero output impedance, and zero noise.

Applications

Use in electronics system design

The use of op-amps as circuit blocks is much easier and clearer than specifying all their individual circuit elements (transistors, resistors, etc.), whether the amplifiers used are integrated or discrete. In the first approximation op-amps can be used as if they were ideal differential gain blocks; at a later stage limits can be placed on the acceptable range of parameters for each op-amp.

Circuit design follows the same lines for all electronic circuits. A specification is drawn up governing what the circuit is required to do, with allowable limits. For example, the gain may be required to be 100 times, with a tolerance of 5% but drift of less than 1% in a specified temperature range; the input impedance not less than 1 megohm; etc..

A basic circuit is designed, often with the help of circuit modeling (on a computer). Specific commercially available op-amps and other components are then chosen that meet the design criteria within the specified tolerances at acceptable cost. If not all criteria can be met, the specification may need to be modified.

Basic non-inverting amplifier circuit

The general op-amp has two inputs and one output. The output voltage is a multiple of the difference between the two inputs

$$U_{out} = G(U_+ - U_-)$$

G is the open-loop gain of the op-amp. The inputs are assumed to have very high impedance; negligible current will flow into or out of the inputs. Op-amp outputs have very low source impedance.

If the output is connected to the inverting input (Fig.1.15), after being scaled by a voltage divider $K=R_1/(R_1+R_2)$, then

$$U_+ = U_{in}$$
$$U_- = K U_{out}$$
$$U_{out} = G(U_{in} - K U_{out})$$

Solving for U_{out}/U_{in}, we see that the result is a linear amplifier with gain

$$U_{out}/U_{in} = G/(1+GK)$$

If G is very large, U_{out}/U_{in} comes close to $1/K$, which equals $1+R_2/R_1$.

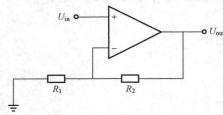

Fig.1.15 Non-inverting amplifier circuit

This negative feedback connection is the most typical use of an op-amp, but many different configurations are possible, making it one of the most versatile of all electronic building blocks [2].

When connected in a negative feedback configuration, the op-amp will try to make U_{out} whatever voltage is necessary to make the input voltages as nearly equal as possible. This, and the high input impedance, are sometimes called the two "golden rules" of op-amp design (for circuits that use negative feedback):

- No current will flow into the inputs;
- The input voltages will be nearly equal.

Most single, dual and quad op-amps available have a standardized pin-out which permits

Chapter 1 Electrical and Electronic Technology Fundamentals

one type to be substituted for another without wiring changes. A specific op-amp may be chosen for its open-loop gain, bandwidth, noise performance, input impedance, power consumption, or a compromise between any of these factors.

Technical Words and Expressions

operational amplifier			运算放大器
DC coupled			直流耦合
differential	[ˌdifəˈrenʃəl]	adj.	差动的
feedback	[ˈfi:dbæk]	n.	反馈
magnitude	[ˈmægnitju:d]	n.	大小
regenerative	[riˈdʒenərətiv]	adj.	再生的
oscillation	[ˌɔsiˈleiʃən]	n.	振荡
array	[əˈrei]	n.	大批；阵列
volume	[ˈvɔlju:m]	n.	量，大量，音量
rugged	[ˈrʌgid]	adj.	坚固的
short circuit			短路
non-inverting input			同相输入
inverting input			反相输入
clarity	[ˈklæriti]	n.	清楚，透明
arithmetic	[əˈriθmətik]	n.	算术，算法
peculiarity	[piˌkju:liˈæriti]	n.	特性
counterpart	[ˈkauntəpɑ:t]	n.	对应物
reliability	[riˌlaiəˈbiliti]	n.	可靠性
voltage amplifier			电压放大器
cascade	[kæsˈkeid]	n.	级联
level shifter			电平转换器
level translator			电平转换
comparator	[ˈkɔmpəreitə]	n.	比较器
hysteresis	[ˌhistəˈri:sis]	n.	迟滞
offset	[ˈɔ:fset]	n.	失调
slew rate			延迟率
tolerance	[ˈtɔlərəns]	n.	误差
drift	[drift]	n.	漂移
versatile	[ˈvə:sətail]	adj.	通用的，万能的
quad	[kwɔd]	n.	4 路
open-loop gain			开环增益
consumption	[kənˈsʌmpʃən]	n.	消耗
compromise	[ˈkɔmprəmaiz]	n.	折中

Notes

[1] Typically the output of the op-amp is controlled either by negative feedback, which largely determines the magnitude of its output voltage gain, or by positive feedback, which facilitates regenerative gain and oscillation.

译文：运放的输出端受负反馈或正反馈控制，负反馈在很大程度上决定输出电压增益的大小，正反馈可以促进再生增益和产生振荡。

注释：which 引导非限制性定语从句。

[2] This negative feedback connection is the most typical use of an op-amp, but many different configurations are possible, making it one of the most versatile of all electronic building blocks.

译文：负反馈连接是运放的最典型应用，但是可能有很多种不同的结构，使得其成为最通用的电子模块之一。

注释：句中 making it one of the most versatile of all electronic building blocks 作句子的补语。

Exercises

Ⅰ. **Mark the following statements with T (true) or F (false) according to the text.**

1. Low input impedance at the input terminals and high output impedance are important typical Characteristics of operational amplifier. ( )
2. Operational amplifier is in the class of nonlinear ICs. ( )
3. The amplifier's differential inputs consist of an inverting input and a non-inverting input, and ideally the op-amp amplifies only the difference in voltage between the two. ( )
4. In the first approximation op-amps can be used as if they were ideal differential gain blocks. ( )
5. The general op-amp has one input and two outputs. ( )

Ⅱ. **Complete the following sentences.**

1. An operational amplifier, often called an op-amp, is a _____high-gain electronic voltage amplifier with _____inputs and, usually, a single output.
2. Input stage consists of high _____it amplifies the _____between the given input signals.
3. In its most common use, the op-amp's _____voltage is controlled by feeding a fraction of the output signal back to the _____input.
4. Unless the differential input voltage is extremely _____, _____operation results in op-amp saturation .
5. A specific op-amp may be chosen for its _____, bandwidth, noise performance, _____, power consumption, or a compromise between any of these factors.

Chapter I Electrical and Electronic Technology Fundamentals

Reading 4 The Basics of Integrated Circuits

Integrated circuits (ICs) are, much as their name would suggest, small circuits integrated into a plastic "chip". They provide a handy source of rich functionality in a tiny package. The ICs used in BEAM bots are really fairly simple —— often just consisting of multiple copies of a simple 2-element or 3-element circuit, in a small, handy, package. This means that they are simple, easy to learn "circuit blocks", rather than very complex "black boxes".

Think of them as just a natural progression from circuits built from discrete components.

ICs come in a variety of packages (2 are predominant for our uses), and are based on a number of different core technologies (also known as logic families, we'll again be concerned with just 2).

Logic families

For the purposes of BEAM bots, we'll just be concerned with two logic families (IC implementation approaches) (Table 1.1); note that both have multiple numbering schemes for their "members".

Table 1.1　　　　　　　　　　　　　　Two logic families

TTL——Transistor/Transistor Logic	TTL chips require a fairly narrow range of supply voltage——5 volts, +/−0.5V. For TTL circuits, a logic "1" is usually defined as a signal of about 2.4V, while logic "0" is about 0.5V above ground; any signal between those values is undefined. Also, note that all unused inputs should be tied to ground. Bipolar TTL logic comes in many flavors (usually numbered in the"74××" format, occasionally in "54××" format) including 74LS/ALS types. These all require 1000s of times more power than TTL-compatible CMOS logic (like the 74HCT type, see below). Note that the 74HCT* ICs have the same logic thresholds as other TTL chips, allowing them used in a mixed circuits of TTL and HCT logic. While we won't use TTL logic in BEAM projects (as a general rule, it's too power-hungry), for completeness, here are some TTL "subfamilies": 74L××——low-power TTL (1/10 the speed, 1/10 the power of "regular" TTL) 74H××——high-speed TTL (twice as fast, twice as much power) 74S××—— Schottky TTL (for high-frequency uses) 74LS××—— combination of low-power & Schottky, same speed as regular TTL, but at 1/5 the power consumption
CMOS——Complimentary Metal Oxide Silicon	CMOS ICs use much less power than TTL, plus they are fairly forgiving of "range" in input voltage —— they're happy with anything between 3 (some, like 1381s, go lower) and 12 volts. CMOS, though, is much more susceptible to damage from static electricity (so get that grounding strap out!). CMOS comes in two flavors, with corresponding numbering schemes: "CD40××" metal gate CMOS, and "74C××" silicon gate CMOS (note that this second subfamily borrows its numbering scheme from TTL ICs). Metal gate CMOS (CD40××) has a rated working voltage of 3~15V, but can be used down to 2V. Silicon gate CMOS (74C××) logic has a working voltage range of 2~6V, but can be used to less than 1V. For microwatt power applications you want to use the lowest possible voltage

Numbering systems and IC "sub-families"

Something I've learned to put up with is the mess that is IC numbering systems. One of the more common numbering systems used for CMOS ICs (most often used in BEAM bots) is borrowed from TTL —— that of 74×××nnn —— here, "×××" is AC, ACT, HC, or HCT (subfamily); and "nnn" is the specific chip ID.

So, basically, any given chip will come in a whole mess of variations (one per subfamily); if it's any consolation, we only need to be concerned with 4. For any given chip, you'll need to consider the plusses and minuses of two comparisons —— AC vs. HC subfamilies, and AC/HC vs. ACT/HCT (essentially, CMOS vs. TTL input levels).

AC vs. HC

AC/ACT stands for Advanced CMOS Logic (ACL for short). HC/HCT stands for high-speed CMOS Logic (HCL).

The AC and ACT subfamilies are faster than the HC and HCT subfamilies, and draw some more power in some circuits. All chips in the AC* subfamily have lower output resistance than HC* and can sink and source 24mA at logic levels and up 70mA (typ) per gate for motor loads. As a result AC* gates can handle more than twice the current of HC* gates (50~70mA vs. 24mA). Note, though, that while most HC* chips have a 25mA limit, the HC* driver chips such as the 74HC240 and the 74HC245 (i.e., buffers) can handle 35mA per device, and a maximum of 75mA per chip.

The AC & ACT families also draw about twice as much current as the HC & HCT chips (but we speaking here of microamps, so it's usually not a huge deal).

It is occasionally possible to find (high quality) motors that you can drive directly from an HC chip. For intermittent operation, such as you get with a quad-core, you could drive very efficient (i.e., very low-current) motors directly. You would definitely need a capacitor (say 0.47 μF) across each motor to keep the noise under control.

CMOS vs. TTL inputs

The main difference between AC-and HC-chips vs. ACT- and HCT-chips is that AC/HC chips work with CMOS-level inputs, while ACT/HCT chips work from TTL-level inputs and outputs. AC/HC chips allow a relatively-wide supply-voltage range (2~6V), and works with CMOS levels on input and output (i.e., doesn't like inputs or outputs to be ill-defined, prefers close to 0V for low and V_{dd} for high). ACT/HCT works only with a supply voltage of 5V ± 10%, and works with TTL levels on input and output (a valid input is close to 0V for low, and 2.4~5.0V for high).

Technical Words and Expressions

progression	n.	进行，前进，进步，发展
BEAM bot		主要方式
flavor	n.	特色

susceptible	adj.	易受影响的
static electricity		静电
numbering scheme		编码方案
subfamily	n.	子系列
consolation	n.	（被）安慰
intermittent	adj.	间歇的，断断续续的
definitely	adv.	明确地

Comprehension

1. For TTL circuits, a logic "1" is usually defined as a signal of about _____, while logic "0" is about _____ above ground.

    A. 3.6V, 1V   B. 5V, 0.7V   C. 2.4V, 0.5V   D. 12V, 1V

2. Note that the 74HCT* ICs have _____ logic thresholds _____ other TTL chips, allowing them used in a mixed circuits of TTL and HCT logic.

    A. the same…as   B. more…than   C. less…than

3. The AC and ACT subfamilies are _____ than the HC and HCT subfamilies, and draw some _____ power in some circuits.

    A. faster, less   B. slower, more   C. faster, more   D. slower, less

4. You would definitely need a _____ across each motor to keep the noise under control.

    A. resistor   B. capacitor   C. diode   D. transistor

5. The main difference between AC- and HC-chips vs. ACT- and HCT-chips is that AC/HC chips work with _____ inputs, while ACT/HCT chips work from _____ inputs and outputs.

    A. CMOS-level, TTL-level   B. CMOS-level, CMOS-level
    C. TTL-level, TTL-level   D. TTL-level, CMOS-level

Unit 5 Digital Logic Gates

Introduction to logic gates

Digital logic gates, which are also known as combinational logic gates or simply "logic gates", are digital ICs whose output at any time is determined by the states of its inputs at that time. Since logic gates are digital ICs, their input and output signals can only be in one of two possible digital states, i.e., logic "0" or logic "1". Thus, the logic state in which the output of a logic gate will be put depends on the logic states of each of its individual inputs.

The primary application of logic gates is to implement "logic" in the flow of digital signals in a digital circuit. Logic in its ordinary sense is defined as a branch of philosophy that deals with what is true and false, based on what other things are true and false. This essentially is the function of logic gates in digital circuits —— to determine which outputs will be true or false, given a set of inputs that can either be true (logic "1") or false (logic "0").

The response output (usually denoted by Q) of a logic gate to any combination of inputs may be tabulated into what is known as a truth table. A truth table shows each possible combination of inputs to a logic gate and the combination's corresponding output. Table 1.2, which describes the various types of logic gates, provides a truth table for each of them as well.

Table 1.2 Logic gates and their properties

Gate	Description	Truth Table		
AND Gate	The AND gate is a logic gate that gives an output of "1" only when all of its inputs are "1". Thus, its output is "0" whenever at least one of its inputs is "0". Mathematically, $Q = A \cdot B$	A	B	Output Q
		0	0	0
		0	1	0
		1	0	0
		1	1	1
OR Gate	The OR gate is a logic gate that gives an output of "0" only when all of its inputs are "0". Thus, its output is "1" whenever at least one of its inputs is "1". Mathematically, $Q = A + B$	A	B	Output Q
		0	0	0
		0	1	1
		1	0	1
		1	1	1
NOT Gate	The NOT gate is a logic gate that gives an output that is opposite the state of its input. Mathematically, $Q = \overline{A}$	A		Output Q
		0		1
		1		0
NAND Gate	The NAND gate is an AND gate with a NOT gate at its end. Thus, for the same combination of inputs, the output of a NAND gate will be opposite that of an AND gate. Mathematically, $Q = \overline{A \cdot B}$	A	B	Output Q
		0	0	1
		0	1	1
		1	0	1
		1	1	0
NOR Gate	The NOR gate is an OR gate with a NOT gate at its end. Thus, for the same combination of inputs, the output of a NOR gate will be opposite that of an OR gate. Mathematically, $Q = \overline{A + B}$	A	B	Output Q
		0	0	1
		0	1	0
		1	0	0
		1	1	0
EXOR Gate	The EXOR gate (for "EXclusive OR" gate) is a logic gate that gives an output of "1" when only one of its inputs is "1"	A	B	Output Q
		0	0	0
		0	1	1
		1	0	1
		1	1	0

Interestingly, the operation of logic gates in relation to one another may be represented and analyzed using a branch of mathematics called Boolean Algebra which, like the common algebra, deals with manipulation of expressions to solve or simplify equations[1]. Expressions used in Boolean Algebra are called, well, Boolean expressions.

Kinds of logic gates

There are several kinds of logic gates, each one of which performs a specific function. These are the: ① AND gate; ② OR gate; ③ NOT gate; ④ NAND gate; ⑤ NOR gate; and ⑥ EXOR gate. Table 1.2 above presents these and their characteristics.

Logic gates may be thought of as a combination of switches. For instance, the AND gate, whose output can only be "1" if all its inputs are "1", may be represented by switches connected in series, with each switch representing an input. All the switches need to be activated and conducting (equivalent to all the inputs of the AND gate being at logic "1"), for current to flow through the circuit load (equivalent to the output of the AND gate being at logic "1").

An OR gate, on the other hand, may be represented by switches connected in parallel, since only one of these parallel switches need to turn on in order to energize the circuit load.

In Boolean Algebra, the AND operation is represented by multiplication, since the only way that the result of multiplication of a combination of 1's and 0's will be equal to "1" if all its inputs are equal to "1". A single "0" among the multipliers will result in a product that's equal to "0". The Boolean expression for "A AND B" is similar to the expression commonly used for multiplication, i.e., A • B.

The OR operation, on the other hand, is represented by addition in Boolean Algebra. This is because the only way to make the result of the addition operation equal to "0" is to make all the inputs equal to "0", which basically describes an "OR" operation. The Boolean expression for "A OR B" is therefore A+B.

The NOT operation is usually denoted by a line above the symbol or expression that is being negated: $\overline{A}$ = NOT(A). The NAND operation is simply an AND operation followed by a NOT operation. The NOR operation is simply an OR operation followed by a NOT operation. The symbols used for logic gates in electronic circuit diagrams are shown in Fig.1.16.

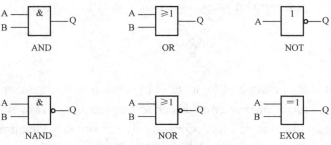

Fig.1.16 Logic gate symbols

One of the most useful theorems used in Boolean Algebra is De Morgan's Theorem, which states how an AND operation can be converted into an OR operation, as long as a NOT operation is available[2]. De Morgan's Theorem is usually expressed in two equations as follows

$$\overline{A \cdot B} = \overline{A} + \overline{B}$$

and

$$\overline{A+B} = \overline{A} \cdot \overline{B}$$

De Morgan's Theorem has a practical implication in digital electronics —— a designer may eliminate the need to add more ICs to the design unnecessarily, simply by substituting gates with the equivalent combination of other gates whenever possible. Since NAND and NOR gates can be used as NOT gates, de Morgan's Theorem basically implies that any Boolean operation may be simulated with nothing but NAND or NOR gates. This is why NAND and NOR gates are also called universal gates.

Technical Words and Expressions

philosophy	[fi'lɔsəfi]	n.	哲学
tabulate	['tæbjuleit]	vt.	把…制成表格
truth table			真值表
AND	[ænd]	n.	与（计算机逻辑运算的一种，或称逻辑乘法）
OR	[ɔ:,ə]	n.	或（逻辑运算的一种）
NOT	[nɔt]	n.	非（逻辑运算的一种）
NAND	[nænd]	n.	与非
NOR	[nɔ:]	n.	或非
Boolean Algebra			布尔代数
manipulation	[mə,nipju'leiʃən]	n.	操作；控制；处理；计算，运算
activate	['æktiveit]	vt.	开（启）动，触发
multiplication	[,mʌltipli'keiʃən]	n.	乘法
multiplier	['mʌltiplaiə]	n.	乘数
negate	[ni'geit]	vt.	求反，非
implication	[,impli'keiʃən]	n.	含意，暗示

Notes

[1] Interestingly, the operation of logic gates in relation to one another may be represented and analyzed using a branch of mathematics called Boolean Algebra which, like the common algebra, deals with manipulation of expressions to solve or simplify equations.

译文：有趣的是，逻辑门与另一个逻辑门之间的运算可以表示出来，并可使用数学的分支——布尔代数来分析。布尔代数，像普通代数那样，可以对要求的表达式进行运算或对等式进行简化。

Chapter 1 Electrical and Electronic Technology Fundamentals

注释：句中 using a branch of mathematics called Boolean Algebra 作状语，which 引导限制性定语从句，修饰 Boolean Algebra。

[2] One of the most useful theorems used in Boolean Algebra is De Morgan's Theorem, which states how an AND operation can be converted into an OR operation, as long as a NOT operation is available.

译文：布尔代数中最有用的定理之一为摩根定理，该定理表达了运用非运算，如何将与运算转化为或运算。

注释：句中 as long as 引导条件状语从句。

Exercises

Ⅰ. Mark the following statements with T (true) or F (false) according to the text.

1. The primary application of logic gates is to implement "logic" in the flow of digital signals in a digital circuit. (　　)
2. A truth table shows some important combinations of inputs to a logic gate and the combination's corresponding output. (　　)
3. An AND gate, on the other hand, may be represented by switches connected in parallel. (　　)
4. The OR operation, on the other hand, is represented by multiplication in Boolean Algebra. (　　)
5. One of the most useful theorems used in Boolean Algebra is De Morgan's Theorem, which states how an AND operation can be converted into an OR operation, as long as a NOT operation is available. (　　)

Ⅱ. Complete the following sentences.

1. The response output (usually denoted by Q) of a logic gate to any combination of inputs may be tabulated into what is known as _____.
2. Expressions used in Boolean Algebra are called, well, _____.
3. Logic gates may be thought of as a combination of _____.
4. The NOR operation is simply an OR operation followed by a _____ operation.
5. A designer may eliminate the need to add more ICs to the design unnecessarily, simply by _____ gates with the _____ combination of other gates whenever possible.

Reading 5　Binary Numbers

Electronic circuits count in binary. This is the simplest possible counting system because it uses just two digits, 0 and 1, exactly like logic signals where 0 represents false and 1 represents true. The terms low and high are also used for 0 and 1 respectively as shown in Table 1.3.

Table 1.3　　　　　　　　　　　　　Logic states

Logic states	
True	False
1	0
High	Low
$+U_s$	0V
On	Off

Counting one, two, three, four, five in binary: 1, 10, 11, 100, 101.

Binary numbers rapidly become very long as the count increases and this makes them difficult for us to read at a glance. Fortunately it is rarely necessary to read more than 4 binary digits at a time in counting circuits.

In a binary number each digit represents a multiple of two (1, 2, 4, 8, 16 etc.), in the same way that each digit in decimal represents a multiple of ten (1, 10, 100, 1000 etc.).

Bits, bytes and nibbles

Each binary digit is called a bit, so 10110110 is an 8-bit number.

A block of 8 bits is called a byte and it can hold a maximum number of 11111111 = 255 in decimal. Computers and PIC microcontrollers work with blocks of 8 bits. Two (or more) bytes make a word, for example PICs work with a 16-bit word (two bytes) which can hold a maximum number of 65535.

A block of 4 bits is called a nibble (half a byte!) and it can hold a maximum number of 1111 = 15 in decimal. Many counting circuits work with blocks of 4 bits because this number of bits is required to count up to 9 in decimal. (The maximum number with 3 bits is only 7).

Hexadecimal (base 16)

Hexadecimal (often just called "hex") is base 16 counting with 16 digits. It starts with the decimal digits 0~9, then continues with letters A (10), B (11), C (12), D (13), E (14) and F (15). Each hexadecimal digit is equivalent to 4 binary digits, making conversion between the two systems relatively easy. You may find hexadecimal used with PICs and computer systems but it is not generally used in simple counting circuits.

Example: 10110110 binary = B6 hexadecimal = 182 decimal.

Technical Words and Expressions

binary	adj. & n.	二进位的，二元的；二进制
counting system		计数系统
decimal	adj. & n.	十进位的；十进制
bit	n.	（二进制）位，比特
byte	n.	（二进制的）字节
nibble	n.	四位字节，半字节

Chapter 1　Electrical and Electronic Technology Fundamentals

block	n.	组
hexadecimal	adj. & n.	十六进制的；十六进制
conversion	n.	转变，变换

Comprehension

1. Electronic circuits count in _____.
   A. Hexadecimal　　B. octal　　C. decimal　　D. binary
2. 110110 in binary equals _____.
   A. 182　　B. 128　　C. 56　　D. 54
3. In a binary number each digit represents a multiple of _____.
   A. two　　B. ten　　C. eight　　D. sixteen
4. Each binary digit is called a _____, a block of 8 _____ is called a _____, A block of 4 _____ is called a _____.
   A. Bit, bits, nibble, bits, byte
   B. Bit, byte, bit, bytes, nibble
   C. Bits, bit, byte, bits, nibble
   D. Bit, bits, byte, bits, nibble
5. _____ (often just called "hex") is base 16 counting with 16 digits.
   A. Hexadecimal　　B. Octal　　C. Decimal　　D. Binary

Unit 6　Flip-flop

In digital circuits, a flip-flop is a term referring to an electronic circuit (a bistable multivibrator) that has two stable states and thereby is capable of serving as one bit of memory. A flip-flop is usually controlled by one or two control signals and/or a gate or clock signal. The output often includes the complement as well as the normal output. As flip-flops are implemented electronically, they require power and ground connections.

Implementation

Flip-flops can be either simple (transparent) or clocked. Simple flip-flops can be built around a pair of cross-coupled inverting elements: vacuum tubes, bipolar transistors, field effect transistors, inverters, and inverting logic gates have all been used in practical circuits —— perhaps augmented by some gating mechanism (an enable/disable input). The more advanced clocked (or non-transparent) devices are specially designed for synchronous (time-discrete) systems; such devices therefore ignore its inputs except at the transition of a dedicated clock signal (known as clocking, pulsing, or strobing). This causes the flip-flop to either change or retain its output signal based upon the values of the input signals at the transition. Some flip-flops change output on the rising edge of the clock, others on the falling edge.

Clocked flip-flops are typically implemented as master-slave devices where two basic flip-flops (plus some additional logic) collaborate to make it insensitive to spikes and noise between the short clock transitions[1]; they nevertheless also often include asynchronous clear or set inputs which may be used to change the current output independent of the clock.

Flip-flops can be further divided into types that have found common applicability in both asynchronous and clocked sequential systems: the SR ("set-reset"), D ("data" or "delay"), T ("toggle"), and JK types are the common ones; all of which may be synthesized from (most) other types by a few logic gates. The behavior of a particular type can be described by what is termed the characteristic equation, which derives the "next" (i.e., after the next clock pulse) output, Q_{next}, in terms of the input signal (s) and/or the current output, Q.

Set-Reset flip-flops (SR flip-flops)

The fundamental latch is the simple SR flip-flop (Fig.1.17), where S and R stand for set and reset respectively. It can be constructed from a pair of cross-coupled NOR logic gates. The stored bit is present on the output marked Q.

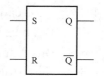

Fig.1.17 The symbol for an SR latch

Normally, in storage mode, the S and R inputs are both low, and feedback maintains the Q and $\overline{Q}$ outputs in a constant state, with $\overline{Q}$ the complement of Q[2]. If S (Set) is pulsed high while R is held low, then the Q output is forced high, and stays high even after S returns low; similarly, if R (Reset) is pulsed high while S is held low, then the Q output is forced low, and stays low even after R returns low (Table 1.4).

Table 1.4 SR flip-flop operation

Characteristic table				Excitation table			
S	R	Action	$Q(t)$	$Q(t+1)$	S	R	Action
0	0	Keep state	0	0	0	×	No change
0	1	Q = 0	0	1	1	0	Set
1	0	Q = 1	1	0	0	1	Reset
1	1	Unstable combination, see race condition	1	1	×	0	No change

("X" denotes a Don't care condition; meaning the signal is irrelevant.)

JK flip-flop

The JK flip-flop (Fig. 1.18) augments the behavior of the SR flip-flop (J=Set, K=Reset) by interpreting the S = R = 1 condition as a "flip" or toggle command. Specifically, the combination J = 1, K = 0 is a command to set the flip-flop; the combination J = 0, K = 1 is a command to reset the flip-flop; and the combination J = K = 1 is a command to toggle the flip-flop, i.e., change its output to the logical complement of its current value. Setting J = K = 0 does NOT result in a D flip-flop, but rather, will hold the current state. To synthesize a D flip-flop, simply set K equal to the complement of J. The JK flip-flop is therefore a universal flip-flop, because it can be configured to work as an SR flip-flop, a D flip-flop, or a T flip-flop. NOTE: The flip flop is positive edge triggered (Clock Pulse) as seen in the timing diagram (Fig.1.19).

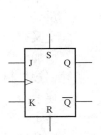

Fig.1.18 A circuit symbol for a JK flip-flop

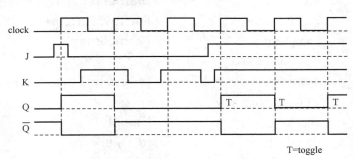

Fig.1.19 JK flip-flop timing diagram

The characteristic equation of the JK flip-flop is

$$Q_{next} = J\overline{Q} + \overline{K}Q$$

and the corresponding truth table is Table 1.5.

Table 1.5   JK flip-flop operation

Characteristic table				Excitation table				
J	K	Q_{next}	Comment	Q	Q_{next}	J	K	Comment
0	0	Q_{prev}	Hold state	0	0	0	×	No change
0	1	0	Reset	0	1	1	×	Set
1	0	Q = 1	Set	1	0	×	1	Reset
1	1	$\overline{Q}_{prev}$	Toggle	1	1	×	0	No change

Uses

- A single flip-flop can be used to store one bit, or binary digit, of data. See preset.
- Any one of the flip-flop types can be used to build any of the others.
- Many logic synthesis tools will not use any other type than D flip-flop and D latch.
- Many FPGA devices contain only edge-triggered D flip-flops.
- The data contained in several flip-flops may represent the state of a sequencer, the value of a counter, an ASCII character in a computer's memory or any other piece of information.
- One use is to build finite state machines from electronic logic. The flip-flops remember the machine's previous state, and digital logic uses that state to calculate the next state.

Flip-flop integrated circuits

Integrated circuits (ICs) exist that provide one or more flip-flops. For example, the 7473 Dual JK Master-Slave flip-flop or the 74374, an octal D flip-flop, in the 7400 series.

Technical Words and Expressions

flip-flop			触发器
bistable	[bai'steibl]	adj.	双稳（态）的
multivibrator	['mʌltivai'breitə]	n.	多谐振荡器
complement	['kɔmplimənt]	n.	反码
cross-coupled			交叉耦合的
augment	[ɔ:g'ment]	v.	增加，增大
gating mechanism			门控机制
enable	[i'neibl]	v.	使…能
disable	[dis'eibl]	v.	禁止
rising edge			上升沿
falling edge			下降沿
clear	[kliə]	adj.	清零
set	[set]	n. & v.	置位
reset	['ri:set]	v.	复位
toggle	['tɔgl]	n.	转换
synthesize	['sinθisaiz]	v.	综合，合成
latch	[lætʃ]	n.	锁存器
flip	[flip]	n.	翻转
command	[kə'mɑ:nd]	n.	指令
trigger	['trigə]	vt.	触发
character	['kæriktə]	n.	字符
sequencer	['si:kwənsə]	n.	程序装置
finite state machine			有限状态机
master-slave flip-flop			主从触发器
octal	['ɔktl]	adj.	八进制的

Notes

[1] Clocked flip-flops are typically implemented as master-slave devices where two basic flip-flops (plus some additional logic) collaborate to make it insensitive to spikes and noise between the short clock transitions.

译文：钟控型触发器一般为主从触发器，主从触发器由两个基本触发器组成，使得触发器对短时钟跃变期间所出现的尖峰和噪声不敏感。

注释：where 引导定语从句，to make…在句中作目的状语。

[2] Normally, in storage mode, the S and R inputs are both low, and feedback maintains the Q and outputs in a constant state, with Q̄ the complement of Q.

Chapter I　Electrical and Electronic Technology Fundamentals

译文：通常在存储模式，S 和 R 输入端都为低电平，反馈维持 Q 和 $\overline{Q}$ 输出恒定，$\overline{Q}$ 为 Q 的反码。

注释：该句为并列句，在句中 with $\overline{Q}$ the complement of Q 作句子的状语。

Exercises

Ⅰ. **Mark the following statements with T (true) or F (false) according to the text.**

1. In digital circuits, a flip-flop is a term referring to an electronic circuit (a bistable multivibrator) that has only one stable states and thereby is capable of serving as one bit of memory.　　　　　　　　　　　　　　　　　　　　　　　　　　　　　　　　(　　)

2. Clocked flip-flops are typically implemented as master-slave devices where two basic flip-flops (plus some additional logic) collaborate to make it insensitive to spikes and noise between the short clock transitions.　　　　　　　　　　　　　　　　　　　　　(　　)

3. The fundamental latch is the simple SR flip-flop, where S and R all stand for set.
　　　　　　　　　　　　　　　　　　　　　　　　　　　　　　　　　　　　　(　　)

4. The JK flip-flop augments the behavior of the SR flip-flop (J=Set, K=Reset) by interpreting the S = R = 1 condition as a "flip" or toggle command.　　　　　　　(　　)

5. To synthesize a D flip-flop, simply set K equal to the complement of J.　(　　)

Ⅱ. **Complete the following sentences.**

1. Some flip-flops change output on the rising edge of the clock, others on the _____ edge.

2. similarly, if R (Reset) is pulsed high while S is held low, then the Q output is forced_____, and stays _____ even after R returns low.

3. To synthesize a D flip-flop, simply set K equal to the _____ of J.

4. The flip flop is _____ edge triggered (Clock Pulse) as seen in the timing diagram shown in Fig.1.19.

5. Many logic synthesis tools will not use any other type than _____ and _____.

Reading 6　Digital Counters

A digital counter, or simply counter, is a semiconductor device that is used for counting the number of times that a digital event has occurred. The counter's output is indexed by one LSB every time the counter is clocked.

A simple implementation of a 4-bit counter is shown in Fig.1.20, which consists of 4 stages of cascaded J-K flip-flops. This is a binary counter, since the output is in binary system format, i.e., only two digits are used to represent the count, i.e., "1" and "0". With only 4 bits, it can only count up to "1111", or decimal number 15.

As one can see from Fig.1.20, the J and K inputs of all the flip-flops are tied to "1", so that they will toggle between states every time they are clocked. Also, the output of each flip-flop in the counter is used to clock the next flip-flop. As a result, the succeeding flip-flop toggles between "1" and "0" at only half the frequency as the flip-flop before it.

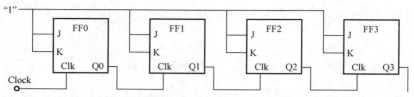

Fig.1.20 A simple ripple counter consisting of J-K flip flops

Thus, in Fig.1.20's 4-bit example, the last flip-flop will only toggle after the first flip-flop has already toggled 8 times. This type of binary counter is known as a "serial", "ripple", or "asynchronous" counter. The name "asynchronous" comes from the fact that this counter's flip-flops are not being clocked at the same time.

A 4-bit counter, which has 16 unique states that it can count through, is also called a modulo-16 counter, or mod-16 counter. By definition, a modulo-k or base-k counter is one that returns to its initial state after k cycles of the input waveform. A counter that has N flip-flops is a modulo 2^N counter.

An asynchronous counter has a serious drawback —— its speed is limited by the cumulative propagation times of the cascaded flip-flops. A counter that has N flip-flops, each of which has a propagation time t, must therefore wait for a duration equal to $N \times t$ before it can undergo another transition clocking.

A better counter, therefore, is one whose flip-flops are clocked at the same time. Such a counter is known as a synchronous counter. A simple 4-bit synchronous counter is shown in Fig.1.21.

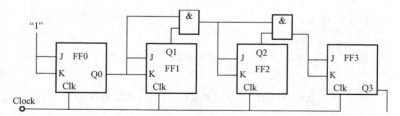

Fig.1.21 A simple synchronous counter consisting of J-K flip flops and AND gates

Not all counters with N flip-flops are designed to go through all its 2^N possible states of count. In fact, digital counters can be used to output decimal numbers by using logic gates to force them to reset when the output becomes equal to decimal 10. Counters used in this manner are said to be in binary-coded decimal (BCD).

Chapter 1 Electrical and Electronic Technology Fundamentals

Technical Words and Expressions

counter	n.	计数器
implementation	n.	执行，履行；落实
stage	n.	[电子学] 级
digit	n.	阿拉伯数字，位数
serial counter		串行计数器
ripple counter		行波计数器
asynchronous counter		异步计数器
modulo-16 counter		16 进制计数器
drawback	n.	缺点，障碍
cumulative	adj.	累积的
propagation	n.	传播，传输
synchronous counter		同步计数器
binary-coded decimal		二-十进制

Comprehension

1. With only 4 bits, a binary counter can only count up to "_____" in binary.
   A. 1110　　　　B. 1111　　　　C. 1011　　　　D. 1001

2. The counter shown in Fig.1.21 is _____ counter.
   A. serial　　　B. ripple　　　C. asynchronous　　　D. synchronous

3. A better counter, therefore, is one whose flip-flops are clocked at the same time. Such a counter is known as a _____ counter.
   A. serial　　　B. ripple　　　C. asynchronous　　　D. synchronous

4. In fact, digital counters can be used to output decimal numbers by using logic gates to force them to reset when the output becomes equal to decimal _____.
   A. 9　　　　　B. 8　　　　　C. 5　　　　　D. 10

Chapter 2

Electric Machines and Motor Control

Chapter 2 Electric Machines and Motor Control

Unit 7 Motor and DC Motor

Fundamentals of motors

An electric motor is a machine used to convert electrical energy to mechanical energy. Electric motors are extremely important to modern-day life, being used in many different places, e.g., vacuum cleaners, dishwashers, computer printers, fax machines, video cassette recorders, machine tools, printing presses, automobiles, subway systems, sewage treatment plants and water pumping stations.

The major physical principles behind the operation of an electric motor are known as Ampere's law and Faraday's law (Fig.2.1). The first states that an electrical conductor sitting in a magnetic field will experience a force if any current flowing through the conductor has a component at right angles to that field[1]. Reversal of either the current or the magnetic field will produce a force acting in the opposite direction. The second principle states that if a conductor is moved through a magnetic field, then any component of motion perpendicular to that field will generate a potential difference between the ends of the conductor.

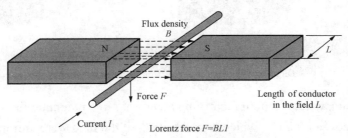

Fig.2.1 Ampere's law and Faraday's law

An electric motor consists of two essential elements. The first, a static component which consists of magnetic materials and electrical conductors to generate magnetic fields of a desired shape, is known as the stator[2]. The second, which also is made from magnetic and electrical conductors to generate shaped magnetic fields which interact with the fields generated by the stator, is known as the rotor. The rotor comprises the moving component of the motor, having a rotating shaft to connect to the machine being driven and some means of maintaining an electrical contact between the rotor and the motor housing (typically, carbon brushes push against slip rings)[3]. In operation, the electrical current supplied to the motor is used to generate magnetic fields in both the rotor and the stator. These fields push against each other with the result that the rotor experiences a torque and consequently rotates.

Electrical motors fall into two broad categories, depending on the type of electrical power applied-direct current (DC) and alternating current (AC) motors.

The first DC electrical motor was demonstrated by Michael Faraday in England in 1821.

Since the only available electrical sources were DC, the first commercially available motors were of the DC type, becoming popular in the 1880s. These motors were used for both low power and high power applications, such as electric street railways. It was not until the 1890s, with the availability of AC electrical power that the AC motor was developed, primarily by the Westinghouse and General Electric corporations.

DC motor

A simple DC motor has a coil of wire that can rotate in a magnetic field. The current in the coil is supplied via two brushes that make moving contact with a split ring. The coil lies in a steady magnetic field. The forces in Fig.2.2 exerted on the current-carrying wires create a torque on the coil.

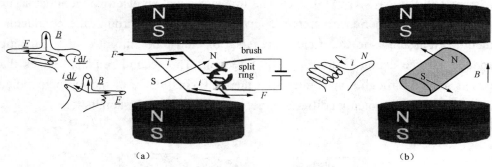

Fig.2.2　Operation principle of DC motors
（a）Direction of F on a wire；（b）Coil indicated by the arrow NS

The force F on a wire of length L carrying a current i in a magnetic field B is iLB times the sine of the angle between B and i, which would be 90°, if the field were uniformly vertical[4]. The direction of F comes from the right hand rule, as shown in Fig. 2.2 (a). The two forces shown in Fig. 2.2 (a) are equal and opposite, but they are displaced vertically, so they exert a torque. (The forces on the other two sides of the coil act along the same line and so exert no torque.)

The coil can also be considered as a magnetic dipole, or a little electromagnet, as indicated by the arrow NS in Fig. 2.2 (b): curl the fingers of your right hand in the direction of the current, and your thumb is the North Pole. In the sketch at right, the electromagnet formed by the coil of the rotor is represented as a permanent magnet, and the same torque (North attracts South) is seen to be that acting to align the central magnet.

Note the effect of the brushes on the split ring. When the plane of the rotating coil reaches horizontal, the brushes will break contact (nothing is lost, because this is the point of zero torque anyway —— the forces act inwards). The angular momentum of the coil carries it past this break point and the current then flows in the opposite direction, which reverses the magnetic dipole. So, after passing the break point, the rotor continues to turn anticlockwise and starts to align in the opposite direction.

The torque generated over a cycle varies with the vertical separation of the two forces. It therefore depends on the sine of the angle between the axis of the coil and field. However, because of the split ring, it is always in the same sense. Fig.2.3 shows its variation in time.

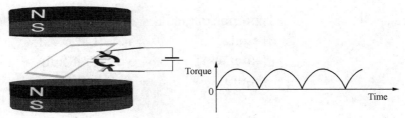

Fig.2.3　Torque variation of DC motor in time

Types of DC motor

DC motors are more common than we may think. A car may have as many as 20 DC motors to drive fans, seats, and windows. They come in three different types, classified according to the electrical circuit used. In the shunt motor, the armature and field windings are connected in parallel, and so the currents through each are relatively independent. The current through the field winding can be controlled with a field rheostat (variable resistor), thus allowing a wide variation in the motor speed. This type of motor is used for driving machine tools or fans, which require a wide range of speeds.

In the series motor, the field winding is connected in series with the armature winding, resulting in a very high starting torque since both the armature current and field strength run at their maximum. However, once the armature starts to rotate, the counter EMF reduces the current in the circuit, thus reducing the field strength. The series motor is used where a large starting torque is required, such as in automobile starter motors, cranes, and hoists.

The compound motor is a combination of the series and shunt motors, having parallel and series field windings. This type of motor has a high starting torque and the ability to vary the speed and is used in situations requiring both these properties such as punch presses, conveyors and elevators.

Technical Words and Expressions

motor	['məutə]	n.	电动机，发动机
convert	[kən'və:t]	vt.	（使）转变，转换
vacuum cleaner			真空吸尘器
dishwasher	['diʃwɔʃə(r)]	n.	洗碗机
machine tool			机床
sewage treatment plant	['sju(:)idʒ]		污水（处理）厂
water pumping station			水泵站
printing press			印刷机

automobile	['ɔ:təməubi:l]	n.	汽车，小汽车
Ampere's law /Faraday's law			安培定律/法拉第定律
magnetic field			磁场
perpendicular	[,pə:pən'dikjulə]	adj.	垂直的，正交的
stator	['steitə]	n.	[机] 定子
interact	[,intər'ækt]	vi.	互相作用，互相影响
rotor	['rəutə]	n.	[机] 转子
rotating shaft			转轴
motor housing			机壳
carbon brush			炭刷
slip ring			滑环
street railway			（市内）电车道
General Electric corporation			通用电气公司 [美]
coil of wire			线圈，线卷，线匝
exert	[ig'zə:t]	vt.	尽（力），施加（压力等）
torque	[tɔ:k]	n.	扭矩，转矩
vertical	['və:tikəl]	adj.	垂直的，直立的
right hand rule			右手定则
magnetic dipole			磁偶极子
electromagnet	[ilektrəu'mægnit]	n.	电磁体，电磁铁
permanent magnet			[物] 永久磁铁
horizontal	[,hɔri'zɔntl]	adj.	地平线的，水平的
angular momentum			角动量
shunt motor			并励 [并绕，分励] 电动机
armature	['ɑ:mətjuə]	n.	电枢，（电动机、发电机的）转子
field winding			励磁绕组
field rheostat			励磁变阻器
series motor			串励 [绕] 电动机
starting torque			启动力矩，启动转矩
armature current			电枢电流
compound motor			复励电机
punch press			冲床

Notes

[1] The first states that an electrical conductor sitting in a magnetic field will experience a force if any current flowing through the conductor has a component at right angles to that field.

Chapter 2 Electric Machines and Motor Control

译文：安培定律指出：对于处在磁场中的带电导体，如果流过导体的电流的某一分量与磁场成直角，那么该导体将受到力的作用。

注释：这是一个宾语从句：the first 指代 Ampere's law，为主语；state 在此取"陈述，规定，指出"的意思，为谓语；that 引导了宾语从句。其中 sitting in a magnetic field 和 flowing through the conductor 为现在分词作定语分别修饰 conductor 和 current。

[2] The first, a static component which consists of magnetic materials and electrical conductors to generate magnetic fields of a desired shape, is known as the stator.

译文：第一部分被称为定子，是由导磁材料和导线组成的静止部件，用来产生所需形状的磁场的。

注释：该句的主干是"The first（指代第一部分）is known as the stator"；中间部分"a static component"与"The first"并列，which 引导了一个定语从句。

[3] The rotor comprises the moving component of the motor, having a rotating shaft to connect to the machine being driven and some means of maintaining an electrical contact between the rotor and the motor housing (typically, carbon brushes push against slip rings).

译文：转子为电动机中的运动部分，转子有一根转轴与被驱动机械相连，并有一些装置维持转子与电动机外壳间的电气接触（通常是炭刷推动滑环）。

注释："having a rotating shaft… and some means…"在句中作状语。

[4] The force F on a wire of length L carrying a current i in a magnetic field B is iLB times the sine of the angle between B and i, which would be 90°，if the field were uniformly vertical.

译文：长为 L 且所带电流为 i 的导线在磁场强度为 B 的磁场中所受的力 F 等于 iLB 乘上 B 与 i 间夹角的正弦，如果磁场是均匀垂线，那么 B 与 i 夹角为 90°。

注释：主语为"The force F on a wire of length L carrying a current i in a magnetic field B"。从句"which would be 90°，if the field were uniformly vertical"为 the angle 的非限定性定语从句，该从句中又包含了一个 if 引导的条件状语从句。

Exercises

I. **Mark the following statements with T (true) or F (false) according to the text.**

1. An electric motor is a machine used to convert mechanical energy to electrical energy.
（ ）

2. Faraday's law states that an electrical conductor sitting in a magnetic field will experience a force if any current flowing through the conductor has a component at right angles to that field.
（ ）

3. The force F on a wire of length L carrying a current i in a magnetic field B is iLB.
（ ）

4. The first DC electrical motor was demonstrated by Ampere in England in 1821. （ ）

5. In the series motor, the armature winding is connected in series with the field winding.
（ ）

II. **Complete the following sentences**

1. The major physical principles behind the operation of an electric motor are known as _____ and _____.

2. An electric motor consists of two essential elements. The first, a static component which consists of magnetic materials and electrical conductors to generate _____ of a desired shape, is known as the _____. The second is known as the _____.

3. A simple DC motor has _____ that can rotate in a magnetic field. The current in the coil is supplied via two _____ that make moving contact with _____. The coil lies in a steady _____. The forces exerted on the current-carrying wires create a _____ on the coil.

4. When the plane of the _____ coil reaches horizontal, the brushes will _____ (nothing is lost, because this is the point of _____ anyway—the forces act inwards). The _____ of the coil carries it past this break point and the current then flows in the opposite direction, which reverses the _____.

5. In the _____, the armature and field windings are connected in _____, and so the currents through each are relatively independent.

Reading 7 Brushless DC Motor

Instead of having the magnets on the stationary casing and the windings on the rotor, we could put the magnets on the rotor and the windings on the stator. That way, we won't need brushes at all because the winding is stationary. However, now we need to find a way to switch the current through the windings at the right moment to ensure the torque on the rotor is always in the same direction. In a conventional motor, this happens automatically as the commutator acts as mechanical switch. With a brushless motor, we need some way to sense the position of the rotor, and then electronically switch the current so it's going the right way through the right winding.

Brushless motors are found in computer hard drives, CD and DVD players, and in anything else where efficiency and reliability are more important than price. As the cost of electronics continues to come down, perhaps one day all DC motors will be built this way.

Advantages: no brushes, simple, efficient, windings are attached to the casing, and easier to cool.

Disadvantages: requires complex drive electronics.

In fact, brushes are bad news. True, they're a clever way to ensure that, as the rotor turns, the current is automatically switched around the windings to ensure the motor keeps turning. However, everything else about them is bad: they are noisy, create friction, generate electrical interference (because of the sparking) and reduce efficiency (because there will always be a voltage drop across the brushes). Not only that, but they eventually wear out. With modern electronics, we can instead sense the position of the rotor (for example, with a Hall-effect device), then switch the current with, for example, a MOSFET transistor.

This is a fan (Fig.2.4) that spent most of its life inside a computer keeping the microprocessor cool. It runs off 12 volts DC and has a brushless motor, as it thoughtfully explains with large friendly letters on the label.

As promised, the magnets are on the rotor (with fan blades attached) in a ring around the outside of the hub. By "feeling" them by using a small compass as a probe, we find that there are four poles, running N-S-N-S around the ring.

Fig.2.4   A fan with a brushless motor

The "stator", in the centre, has four small coils with shaped pole pieces to create a strong magnetic field next to the rotor. Depending on which director the current flows through each coil, it will attract or repel a north pole. So, all we have to do is to keep switching the direction of current flow through the coils in synchronization with the rotation of the magnets, and we'll keep exerting a torque that keeps the fan turning.

Now we've peeled the label off and can see the electronics that does the switching. It consists of a single integrated circuit and a few small capacitors, so it's actually not all that complex! If we google the part number of the chip (LB1962M), we find it is a "*Fan motor single phase full-wave driver*", which I guess is reassuring.

But how does the motor know the exact moment that the magnet has passed one pole, and therefore that it's time to reverse the current flow? There are three techniques commonly used:

- Hall-effect sensors. This is a neat, non-contact way of knowing where the magnets are.
- Back EMF. This is even neater. We don't use sensors at all, but use the fact that the magnet moving past the coil will induce a voltage in it, and use this voltage to tell us where the magnet is.
- Don't bother. For the ultimate minimalist approach, just keep switching the coils in sequence and assume the rotor will keep up. For motors with a small load that is well defined (e.g., a fan), this works pretty well.

Technical Words and Expressions

brushless	adj.	无刷的
friction	n.	摩擦，摩擦力
synchronization	n.	同步
reassuring	adj.	可靠的
Hall-effect sensor		霍尔传感器
back EMF		反电动势

Comprehension

1. Perhaps one day all DC motors will be brushless motors, because _____.

A. brushless motors are more efficiency and reliability than conventional motors

B. the cost of electronics continues to come down

C. the cost of brushless motors comes down

D. both A and C

2. which of the following is right? _____.

A. Brushless motors are only found in computer hard drives, CD and DVD players

B. With a brushless motor, we have to measure the position of the rotor

C. With a brushless motor, the magnets are on the stationary casing and the windings are on the rotor

D. Brushless motors always switch direction of the torque on the rotor

3. Brushes are bad news, because _____.

A. they are noisy and create friction, generate electrical interference and increase efficiency

B. they ensure the motor keeps turning

C. one day they will be out of workout

D. all above

4. LB1962M is _____.

A. a "Fan motor there phase full-wave driver"

B. a amplifier used in a fan motor

C. a "Fan motor single phase full-wave driver"

D. a power supply used in a fan motor

5. How does the brushless motor know that it's time to reverse the current flow? _____.

A. Hall-effect sensors

B. Back electromotive force

C. Keep switching the coils in sequence and assume the rotor will keep up

D. All above

Unit 8 Induction Motors

No modern home should be without one —— or maybe a dozen. You'll find an induction motor in the fan, fridge, vacuum cleaner, washing machine, dishwasher, clothes drier, and the little pump that circulates water in the fish tank to stop the water turning green and the fish going belly-up. Chances are there's also one in the air conditioner —— unless it's a particularly high-tech one[1].

Advantages: cheap, quiet, long lasting, creates no interference.

Disadvantages:

- wants to turn at constant speed (50Hz divided by half the number of poles);
- cannot turn faster than 1500r/min (4-pole motor);
- draws a massive starting current, or is inefficient, or both;

- kind of big and bulky for the power it develops.

Induction motor in Fig.2.5 came out of a fan.

Actually, the bearings and end-caps of the motor have already been removed. We can pull the rotor out and Fig.2.6 (b) is what we're left with. There are four windings, and they are all simply in series.

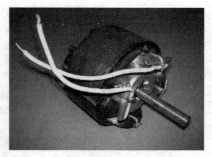

Fig.2.5 Signal phase induction motor

Well, not quite simply —— the current comes in the white wire, then the first winding [top right in Fig.2.6 (a)] is anticlockwise, the next one [bottom right in Fig.2.6 (a)] is clockwise, bottom left [in Fig.2.6 (a)] is anticlockwise again, top left is clockwise, then out the other white wire. So, imagine a positive half-cycle of the mains, with the current actually coming in that first wire. The first winding produces a north pole facing in; the second a south pole facing in; etc., like this: N-S-N-S.

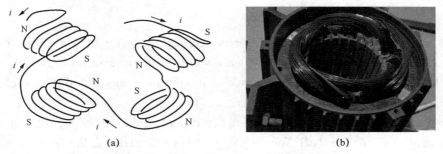

Fig.2.6 Signal phase induction motor without rotor
(a) Current in windings; (b) Four windings in series

Half a mains cycle later (10ms) the current has reversed and so must the magnetic sense of the poles, which are now: S-N-S-N[2]. Now the magnetic field of stator have rotated 90 degrees. The rotor is an electrical conductor, and therefore tries to follow this field. To do so it has to rotate through 90 degrees. The rotor thus takes two full cycles of the mains (40ms) to make a complete rotation, and so revolves at 1500r/min. At least, it would if it could keep up with the rotating field. But it can't, quite, and in fact it's only because it's slipping behind that any torque is developed at all[3]. So, it rotates a bit slower than 1500r/min (typically 1440r/min) depending on how much torque it is being called upon to produce.

Note that the motor, as described so far, could rotate happily clockwise or anticlockwise. This kind of motor therefore needs some kind of internal cleverness to ensure it only turns in the right direction. This is achieved, in this motor, by the use of shaded poles.

Notice the winding at the top of Fig.2.7. See how there is a small additional pole (or set of iron laminations) off to the left of the main pole. It's excited by the same winding as the main pole, but is "shaded" from it by a thick copper band that wraps around the laminations and acts like a shorted electrical turn. The current induced in this band by the magnetic field generates a phase shift so that the shaded pole can generate a small component of magnetic field at right angles to the main field, and with the correct phase to ensure the fan turns the right way (otherwise the fan would suck instead of blowing)[4].

So in fact our induction motor is using induction already, and we haven't even got to the rotor yet!

Now we look at the rotor (Fig.2.8). This is a real disappointment —— it looks nothing like the "squirrel cage" in the text book! Where's the squirrel supposed to go, for starters?

Fig.2.7 Shaded poles   Fig.2.8 Rotor

What's happened here is that the rotor is actually made up of a stack of disc-shaped laminations of soft iron. That's right —— it's solid. This concentrates the magnetic field (generated by the windings) into the region where it will do the most good (the conducting bars of the rotor).

You can actually see the edges of the bars that run along the axis of the rotor, but they're at an angle of maybe 30° to the shaft. What's going on here? Bad day at the factory? Chances are it's been designed that way to reduce cogging torque. If the bars ran parallel to the axis, the torque would rise and fall as each bar passed under the windings. By slanting the bars, the torque is kept more uniform as the rotor turns.

Now let's look at a different type of induction motor (Fig.2.9).

(a)　　　　　　　　　　(b)

Fig.2.9 Induction motor came out of an astronomical telescope
(a) Induction motor; (b) Stator

This induction motor came out of an astronomical telescope. It was part of the photographic film transport, and needed to be able to turn both forwards and backwards. It therefore has two separate windings, and four wires coming out. One winding is fed directly from the mains (or "line" as our US colleagues call it); the other is fed through a capacitor that provides the necessary 90° phase shift. Swap the windings over, or reverse the connections to one of the windings, and the motor goes the other way.

No surprises when we take it apart, although note a very different winding pattern to the previous motor. It has more poles, and therefore turns slower.

Once again, the rotor is solid, and we can't see what's inside. The aluminium plate at the end of the rotor has been stamped and turned up into a series of small fins to make a crude cooling fan (Fig.2.10). (This wasn't necessary with our first motor —— it kept itself cool by the simple expedient of placing itself in the middle of, well, a fan.)

(a) (b)

Fig.2.10 Rotor

Since astronomical telescopes no longer use film, we may as well cut the rotor in half and see if there's a squirrel in there.

No squirrel, but a magnificent set of aluminium conducting bars, just like in the text books. If you think of the rotor bars as forming (via the end rings) a single-turn secondary winding of a transformer, the primary of which (the windings on each pole) has some 50~100turns, it is clear that the current through the rotor bars can be very high —— as much as 100amps for a 240-watt motor. This explains the need for really chunky bars!

One disadvantage of the shaded pole motor is that the starting torque is rather low. This doesn't matter for something like a fan, where the load when stationary is almost zero. For other applications, like a washing machine, it would be a disaster. Such motors therefore use a capacitor to generate the required phase shift for the quadrature windings, as in this example.

Induction motors also come in other variations, but the two described above are the most common in domestic use.

For serious grunt, however, you need a three phase induction motor. This takes advantage of the fact that commercial 3 phase power is delivered by three conductors, each of which carries a 50Hz sine wave with 120° of phase shift relative to the other two. A 3 phase motor simply places three windings at 120° intervals around the casing, and a rotating magnetic field is automatically produced. Three phase induction motors are the "workhorse" of industry, with large units having ratings well in excess of a megawatt.

Technical Words and Expressions

induction motor			感应电动机
bearing	[ˈbɛəriŋ]	n.	轴承
end-cap		n.	弹簧盒盖，端盖
disassemble	[ˌdisəˈsembl]	vt.	解开，分解
r/min			转/分钟
internal	[inˈtəːnl]	adj.	内在的，内部的
lamination	[ˌlæmiˈneiʃən]	n.	叠片结构
copper band			紫铜带/片
squirrel cage			鼠笼
a stack of			一堆
soft iron			软铁
concentrate	[ˈkɔnsentreit]	v.	集中，浓缩
cogging torque			齿槽效应转矩，磁阻转矩，齿滞
astronomical telescope			天文望远镜
photographic film transport			摄影胶片输片机构
main	[mein]	n.	（供电）干线
phase shift			相移
swap	[swɔp]	n.&v.	交换
aluminium plate			铝板，铝片
rotor bar			转子铜条，转子导条
secondary winding			二次绕组，次级绕组
shaded pole motor			罩极电动机
starting torque			启动转矩
quadrature winding			正交磁绕组，正交绕组，交磁绕组
three-phase induction motor			三相感应电动机
workhorse	[ˈwɜːkhɔːs]	n.	重负荷机器
megawatt	[ˈmegəwɔt]	n.	兆瓦特

Notes

[1] Chances are there's also one in the air conditioner —— unless it's a particularly high-tech one.

译文：空调中可能也会有一个感应电动机——除非它是特殊的高科技空调。

注释：The chance is (that) .../The chances are (that) ...是固定搭配，表示"有可能……"。

[2] Half a mains cycle later (10ms) the current has reversed and so must the magnetic sense of the poles, which are now: S-N-S-N.

译文：半个电源周期（10ms）后电流反向，因此磁场的极向也反向，现在磁极方向为：S-N-S-N。

注释："so must the magnetic sense of the poles" 为倒装句型，一般以 so, nor, neither 开头，表示谓语所述的情况也适用于另一个人或一事物的肯定或否定句中，需要倒装，so 用于肯定句，表示"也一样""也这样"；nor, neither 用于否定句，表示"同样也不，也不这样"。该句正常语序为 "the magnetic sense of the poles must have reversed"。

[3] But it can't, quite, and in fact it's only because it's slipping behind that any torque is developed at all.

译文：但是转子和旋转磁场的转速不完全一致，事实上正是因为转子转速比旋转磁场低，才使得转矩产生。

注释："it can't, quite"是省略句式，完整的句子为"it can't keep up with the rotating field"。

[4] The current induced in this band by the magnetic field generates a phase shift so that the shaded pole can generate a small component of magnetic field at right angles to the main field, and with the correct phase to ensure the fan turns the right way (otherwise the fan would suck instead of blowing).

译文：磁场在铜片中感应出电流，这个电流产生一个相位移，因此罩极能够给主磁场增加一个向右的小磁场分量，由于有正确的相位，可确保风扇按正确的方式旋转（否则风扇将吸风，而不是吹风）。

注释：并列连词 and 引导两个因果关系的从句。"and"前的句子结构："so that 因此，所以"，引导一个表示结果的状语从句，其主句结构为"[The current (induced in this band by the magnetic field 过去分词做定语)]（主语）＋generates（谓语）＋a phase shift（宾语）"。

Exercises

Ⅰ. Mark the following statements with T (true) or F (false) according to the text.

1. There are four windings in two-phase induction motor , and they are all simply in parallel.　　　　　　　　　　　　　　　　　　　　　　　　　　　　　（　）
2. The rotor is an electrical conductor, and therefore tries to follow rotor magnetic field. （　）
3. The rotor of induction motor is actually made up of a stack of disc-shaped laminations of soft iron.　　　　　　　　　　　　　　　　　　　　　　　　　　　　（　）
4. One advantage of the shaded pole motor is that the starting torque is rather low.　（　）
5. For serious grunt, however, you need a three-phase induction motor.　　　（　）

Ⅱ. Complete the following sentences

1. The disadvantages of induction motor: Wants to turn at _____(50Hz divided by half the number of poles)；Cannot turn faster than 1500rpm (4-pole motor)；Draws a massive _____, or is _____, or both.

2. You can actually see the edges of the bars that run along the axis of the rotor, but they're at an angle of maybe 30 degrees to _____. It's been designed that way to reduce _____. If the _____ran parallel to the axis, the torque would rise and fall as each bar passed under _____.

3. The rotor is _____, and we can't see that the _____at the end of the rotor has been stamped and turned up into a series of small fins to make _____.

4. For other applications, like a washing machine, it use _____ to generate the required _____ for the _____.

5. A 3-phase motor simply places _____ at 120 degree intervals around the casing, and _____ is automatically produced.

Reading 8 Synchronous Machines: Generalities

The synchronous machine shown in Fig.2.11 is a particular type of rotating electrical machine having a rotation speed strictly connected to the frequency of the sinusoidal electrical quantities at the terminals.

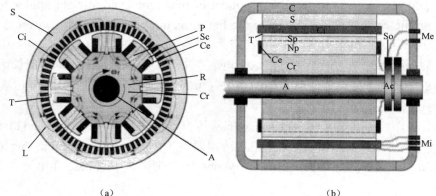

Fig.2.11 Synchronous machine
(a) Cross section; (b) Longitudinal section
S—stator; P—salient pole; Ci—slots and armature conductors;
Sc—pole shoe; T—air gap; Ce—excitation conductors; L—induction vectorial lines;
R—rotor; A—shaft; Cr—rotor crown; C—frame; Ac—conducting rings; So—brushes;
Sp—pole shoe; Me—excitation terminals; Np—polar core; Mi—armature terminals

In the operation as a generator it is called alternator; when it operates as a motor it is called synchronous motor.

There are no fundamental differences between the two conditions and often the same machine can operate both as an alternator and as a motor. Even if the motor has important uses, in any case the fundamental application of the synchronous machine is as an alternator. As a matter of fact, nearly the whole of the electrical power used in the world is produced with alternators.

Usually the inductor is set in the rotor and the armature in the stator.

Rotor

The rotor can have salient poles, generally for a number of polar couples $p > 3$, or it can be smooth (generally for $p < 2$). In the first case the rotor includes a massive steel crown with polygonal section integral to the shaft. On it we have mounted the $2p$ polar cores in massive steel, and the pole shoes, that can be of massive steel or with bars, because their induction has time fluctuations. Around the polar cores the inductor or excitation conductors are wound, to form $2p$ coils identical between them.

The smooth rotor has been carried out in massive steel and it is equipped with longitudinal slots housing the inductor or excitation conductors; the slot distribution and the connections between the conductors are such to carry out two (or four) identical coils.

The coils of the consecutive poles are always connected in antiseries and the inductor winding resulting from it is supplied through creeping contacts made up of brushes and rings; the constant excitation current I_e is provided by a specific d.c. generator (exciter), that is often made up of a rotating generator, keyed on the same shaft of the synchronous machine.

Sometimes, in machines having not a very big power, the poles are magnetized by means of permanent magnets instead of excitation coils. In this case, being the excitation current not required, the machines are without rings and brushes and they are called brushless.

There are also synchronous machines (of small power) where the inductor is set in the stator, which is therefore equipped with salient poles, excited by coils or permanent magnets.

Stator

The stator, having the shape of a cylindrical crown, is carried out in bar iron, being the seat of periodically variable magnetic induction.

On its internal edge there are longitudinal slots of length L, evenly spaced, housing the armature conductors.

These are connected to carry out the armature windings that, connected among them, build up one or, more often, three armature windings, according to the fact that the machine is single phase or three-phase. The winding terminals are connected to the main terminals of the machine, and later they have to be connected to a mains in single phase or three-phase.

Inductor and armature conductors

The armature and inductor conductors are generally made of copper, isolated among them and as to the iron parts.

In the inductor (operating in d.c.) conductors of high section can be used; on the contrary in the armature (operating in a.c.), to reduce the skin effect, we apply to thinner elementary conductors, reciprocally isolated, that are transposed among them and parallel connected to the edges.

Similarly to the transformer ones, the rated values of a synchronous machine locate the levels of the main quantities with which we obtain the optimum operation of the machine.

The main rated values of the synchronous machine are listed:

P_n Rated power [VA, W]
U_n Armature rated voltage [V]
I_n Armature rated current [A]
f_n Rated frequency [Hz]
I_e Excitation current [A]
U_e Excitation voltage [V]
n_n Rotation rated speed [r/min]
C_n Rated couple at the shaft [Nm]

Armature rated voltage and current have the meaning of effective values.

In the case of alternators the rated power has the meaning of apparent delivered electrical power and it is expressed in VA; In the case of motors it has on the contrary a meaning of power returned to the shaft and it is expressed in W.

For single phase and three-phase alternators the following relationships are respectively valid

$$P_n = U_n I_n, \quad P_n = \sqrt{3} U_n I_n$$

On the contrary for the synchronous motors we have

$$P_n = C_n \frac{2\pi}{60} n_n$$

Technical Words and Expressions

rotating electrical machine		旋转电机
alternator	n.	交流发电机
synchronous motor		同步电动机
salient pole		凸极
air gap		气隙
pole shoe		极靴
polygonal	adj.	多角形的，多边形的
fluctuation	n.	波动，起伏
longitudinal slot		纵向槽
rated	adj.	额定的
excitation current		励磁电流

Comprehension

1. Which of the following is right? _____.

   A. Alternator is a kind of synchronous machine

   B. Alternator is a kind of rotating electrical machine

   C. Alternator is synchronous motor

D. both A and B
2. Compared to synchronous motor, _____.
   A. synchronous generator's fundamental is more complex
   B. synchronous generator's construct is more complex
   C. synchronous generator's applications are wider
   D. all above
3. Which of the following states about rotor of synchronous machine is right? _____.
   A. If the rotor is smooth, it includes a massive steel crown with polygonal section integral to the shaft
   B. Rotor's excitation current I_e is provided by a specific a.c. generator
   C. The inductor of synchronous machines can't be set in the stator
   D. If the synchronous machines don't have a very big power, the rotor can be made of permanent magnets
4. Which of the following states about stator of synchronous machine is wrong? _____.
   A. The stator looks like a cylindrical crown
   B. The stator is located in a periodically variable magnetic filed
   C. On its internal edge there are longitudinal slots of length, housing the armature conductors
   D. Each winding's terminals are directly connected to the mains terminal
5. Which of the following states about synchronous generator is right? _____.
   A. The inductor can use thinner section conductors
   B. Armature rated voltage and current are effective values
   C. The rated power has the a meaning of power returned to the shaft and it is expressed in VA
   D. The armature uses high section elementary conductors to reduce the skin effect

Unit 9 Simple Motor Control Circuits

This article will discuss some simple AC motor control circuits.

Button lock circuit

The circuit shown in Fig.2.12 shows how to use contactor to lock button position. When "START" button is pressed "KM" activates and motor starts to rotate. At the same time one contact of contactor "KM" locks start button so that current continues to pass through "KM". Stop button is used to cut the energy from contactor "KM", which will cut energy from motor.

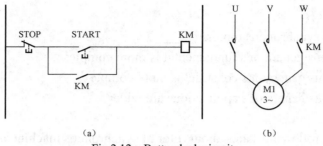

Fig.2.12　Button lock circuit
(a) Command circuit; (b) Motor circuit

Remote control circuit

The circuit shown in Fig.2.13 is used to control motor from different locations. All stop buttons are connected serially and start buttons connected parallel to each other. We can start and stop motor by using any start and stop buttons from any location.

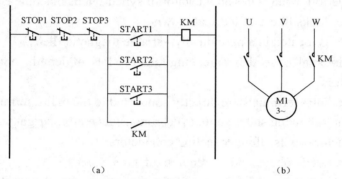

Fig.2.13　Remote control circuit
(a) Command circuit; (b) Motor circuit

Starter circuit for one phase asynchronous motor

The circuit shown in Fig.2.14 is used to start one phase asynchronous motor. By pressing start button "KA" and "KM" are activated and motor starts to rotate[1]. At the same

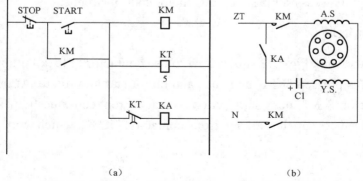

Fig.2.14　Starter circuit for one phase asynchronous motor
(a) Command circuit; (b) Motor circuit

time, timer "KT" activates and starts to count. When timer finishes counting, "KA" is cut off from energy by the contactor of time relay and motor continues to rotate without auxiliary winding[2].

Motor starter circuit with single resistance

When the stator windings of an induction motor are connected directly to its 3-phase supply, a very large current (5~8 times full load current) flows initially. This surge current reduces as the motor accelerates up to its running speed. When very large motors are started direct-on-line they cause a disturbance of voltage (voltage dip) on the supply lines due to the large starting current surge[3]. This voltage disturbance may result in the malfunction of other electrical equipment connected to the supply. To limit the starting current some large induction motors are started at reduced voltage and then have the full supply voltage reconnected when they have run up to near rated speed[4].

The circuit shown in Fig.2.15 is used to start high power induction motors. When we click on start button "KM1" and "KT" time relay are activated. Contactors of relay "KM1" connect motor to 3 phase power supply and auxiliary contact is used to keep relay in power. The motor voltage is reduced by resistances serially connected to the motor. Time relay "KT" starts to count down. When time relay finishes counting, "KM2" is activated by contactor of time relay. Contactors of "KM2" bypass resistances used in the motor circuit and motor continues rotating with full supply voltage. Timer of time relay "KT" must be arranged to count until motor gets to rated speed. Button Stop is used to cut power off from motor and all contactor used in command circuit.

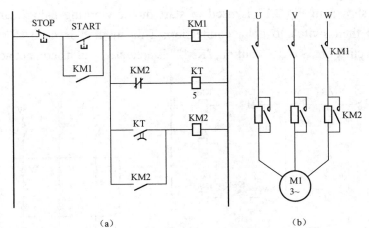

Fig.2.15 Motor starter circuit with single resistance
(a) Command circuit; (b) Motor circuit

Motor starter circuit with double resistance

High power motors always need starter circuit for starting to work. In Fig.2.16, when we click on start button motor starts to work with low voltage because of usage of two resistances. First timer starts to count to specified time. When timer finishes counting "KM2" activates which is used to bypass first resistance in motor circuit and starts second timer. When second timer finishes counting "KM3" activates and both resistances are bypassed in motor circuit.

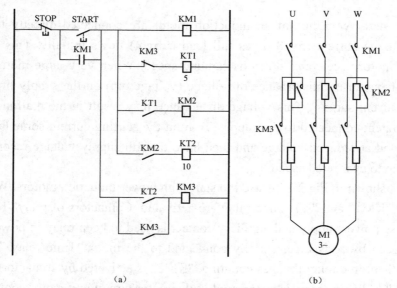

Fig.2.16 Motor starter circuit with double resistance
(a) Command circuit; (b) Motor circuit

Star-Delta starter circuit

The circuit shown in Fig.2.17 is used to start motor working using star connection for some period and then switch to delta connection. This method is used to start high power motors. When we click on "START" button, "KM2" is activated and its contact activates "KM1",

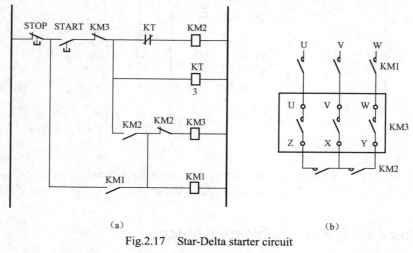

Fig.2.17 Star-Delta starter circuit
(a) Command circuit; (b) Motor circuit

which connects motor to power supply. Other contacts of "KM2" connect motor windings in star connection. Motor starts to rotate using low voltage. Time relay "KT" is also activated by "START" button. When time relay finishes counting its normally closed contact activates and "KM2" is deactivated, which activates "KM3", "KM3" converts motor connection type to delta. Auxiliary contact of "KM2" deactivates time relay "KT". When we click on "STOP" button all contactors are deactivated and motor circuit is cut off from power supply.

Changing rotation direction with protection lock

In Fig.2.18, we have two buttons to rotate motor in separate directions. "SB1" button activates "KM1" which rotates motor in forward direction and "SB2" button activates "KM2" which rotates motor in backward direction. At the same time button "SB1" is used to cut energy from "KM2", in case it was activated, before activating "KM1". This is preventing from short circuit in motor circuit. Button "SB2" is also used in the same way. "STOP" button is used to cut energy from both "KM1" and "KM2", which cuts energy from motor.

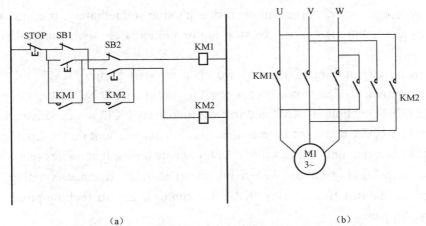

Fig.2.18 Changing rotation direction with protection lock
(a) Command circuit; (b) Motor circuit

Changing rotation direction with electric lock

In Fig.2.19, to rotate the motor in selected direction press "SB1" or "SB2" buttons. While the motor is rotating in one direction it is impossible to rotate it in another direction without stopping it[5]. If you try to change direction of rotation while it is rotating in one direction. Command circuit will not activate related relay because of security contact. For example we start to rotate the motor in forward direction by pressing "SB1" button."KM1" is activated and the motor starts to rotate. Normally closed contact of "KM1" located before "KM2" will prevent activation of "KM2". Therefore we have to stop the motor by clicking "STOP" button and deactivate active "KM1" in case to rotate the motor in reverse direction.

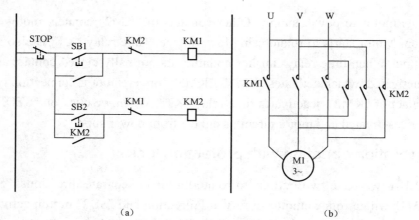

Fig.2.19 Changing rotation direction with electric lock
(a) Command circuit; (b) Motor circuit

Dynamic forced braking

When we cut off energy from motor it doesn't stop immediately. It continues rotating until it looses all its kinetic energy. To stop motor immediately dynamic braking circuit is used.

In command circuit (Fig.2.20) when we click on start button "KM" and inverse time relay "KT" are activated. The motor is connected to power supply by contacts of relay "KM". When we click "STOP" button "KM" and inverse time relay "KT" are deactivated. The motor is cut off from energy. Counter of inverse time relay starts to count down. Contact of inverse time relay "KT" is remaining active while counter counts down. It activates relay "KA" which connects motor to braking circuit. When inverse time relay finishes counting, its contact became inactive, that deactivates relay "KA". The motor is cut off braking power supply and remains fully stopped.

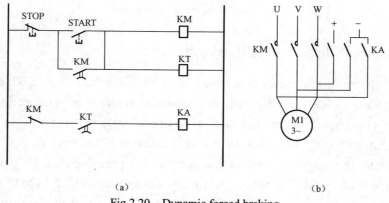

Fig.2.20 Dynamic forced braking
(a) Command circuit; (b) Motor circuit

Optional dynamic braking

The circuit shown in Fig.2.21 has optional dynamic braking button "FREN". Start motor rotation by pressing "START" button. To stop it we have two choices. First by pressing "FREN" button which will cut main power and apply braking circuit to the motor for short period. Second by pressing "STOP" button which will just cut power from motor circuit.

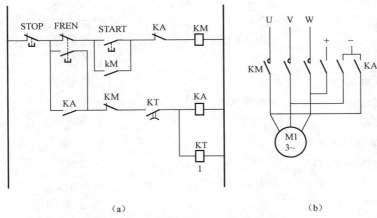

Fig.2.21 Optional dynamic braking
(a) Command circuit; (b) Motor circuit

Continuous and discontinuous working motor circuit

The circuit shown in Fig.2.22 explains how control the motor for continuous or discontinuous working. By pressing "START" button we make the motor work continuously. "KM" is used to rotate the motor and lock start button. By pressing "KS.CL" button we use "KM" just for rotating the motor. When we release "KS.CL" button "KM" energy will be cut off and the motor will stop. "STOP" button is used to stop the motor in continuous working mode.

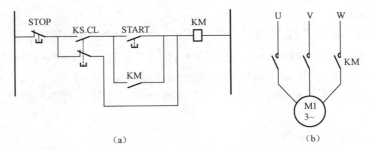

Fig.2.22 Continuous and discontinuous working motor circuit
(a) Command circuit; (b) Motor circuit

Technical Words and Expressions

contact	[ˈkɔntækt]	n.	（机器的）触点、接头、接触器
button	[ˈbʌtn]	n.	按钮
timer	[ˈtaimə]	n.	定时器
auxiliary winding			辅助线圈
surge current			冲击（浪涌）电流
disturbance	[disˈtə:bəns]	n.	干扰，骚乱，搅动
voltage dip			电压骤降
starting current surge			启动电流浪涌
malfunction	[mælˈfʌŋkʃən]	n.	故障
auxiliary contact			辅助触点；联锁触点
Star-Delta starter circuit			星-三角启动器电路
normally closed contact			常闭触点（动断触点）
deactivate	[di:ˈæktiveit]	vt.	失电
in forward direction			［电］正向
kinetic energy			动能
dynamic braking			能耗制动

Notes

[1] By pressing start button "KA" and "KM" are activated and motor starts to rotate.

译文：按下启动按钮，"KA"和"KM"得电，电动机开始旋转。

注释：此句中"and"连接了三个并列句，这三个句子间有因果次序关系。

[2] When timer finishes counting, "KA" is cut off from energy by the contactor of time relay and motor continues to rotate without auxiliary winding.

译文：当计数结束后，时间继电器的触点动作使"KA"断电，从而电动机在没有辅助线圈的情况下继续转动。

注释：此句为"when"引导的时间状语从句，主句为"and"连接了两个并列句。

[3] When very large motors are started direct-on-line they cause a disturbance of voltage (voltage dip) on the supply lines due to the large starting current surge.

译文：当超大型电动机直接启动时，这个大启动冲击电流，将会给供电线路带来电压扰动（电压骤降）。

注释：when 引导状语从句。

[4] To limit the starting current some large induction motors are started at reduced voltage and then have the full supply voltage reconnected when they have run up to near rated speed.

译文：为了限制启动电流，某些大型的感应电动机采用降压启动，当电动机接近额定转速后再给电动机加全压。

Chapter 2 Electric Machines and Motor Control

注释："To limit the starting current"表示目的，该句为"and"连接的两个并列句，其中第二个并列句带一个"when"引导的时间状语从句。

[5] While the motor is rotating in one direction it is impossible to rotate it in another direction without stopping it.

译文：当电动机在一个方向上转动时，除非让它停下来，否则不可能在另一个方向上转动。

注释："While"引倒条件状语从句，从句为"the motor is rotating in one direction"，主句为"it is impossible to…without stopping it"。"without"表示"除非"的意思。

Exercises

I. Mark the following statements with T (true) or F (false) according to the text.

1. Button lock circuit is used to control motor from different locations.　　(　)
2. When very large motors are started direct-on-line they cause a disturbance of voltage (voltage dip) on the supply lines due to the large running current surge.　　(　)
3. High power motors always need starter circuit for starting to work.　　(　)
4. Star-Delta starter circuit is used to start motor working using delta connection for some period and then switch to star connection.　　(　)
5. When we cut off energy from motor it doesn't stop immediately.　　(　)

II. Complete the following sentences.

1. In Remote control circuit all _____ connected _____ and start buttons connected _____ each other. We can start and stop motor by using any start and stop buttons from _____.

2. When the _____ of an induction motor are connected directly to its _____, a very large current (5-8 times _____) flows initially. This surge current reduces as the motor accelerates up to its _____.

3. To limit the starting current some large induction motors are started at _____ and then have the _____ reconnected when they have run up to near _____.

4. When we click on start button in Motor Starter circuit with single resistance, "M" _____ and "ZR" _____ _____. _____ of relay "M" connect motor to 3 phase power supply and _____ used to keep relay in power.

5. To stop motor immediately _____ circuit is used.

Reading 9 How to Wire a Motor Starter

A motor starter is a combination of devices used to start, run, and stop an induction motor based on commands from an operator or a controller. In North America, an induction motor will typically operate at 230V or 460V, 3-phase, 60Hz and has a control voltage of 115V AC or 24 VDC. Several other combinations are possible in North America and other countries, and are easily derived from the methods shown in this document.

The motor starter must have at least two components to operate: a contactor to open or close the flow of energy to the motor, and an overload relay to protect the motor against thermal overload. Other devices for disconnecting and short-circuit protection may be needed, typically a circuit breaker or fuses. Short-circuit protection will not be shown in the examples that follow.

The contactor is a 3-pole electromechanical switch whose contacts are closed by applying voltage to a coil. When the coil is energized, the contacts are closed, and remain closed, until the coil is de-energized. The contactor is specifically designed for motor control, but can be used for other purposes such as resistive and lighting loads. Since a motor has inductance, the breaking of the current is more difficult so the contactor has both a horsepower and current rating that needs to be adhered to.

The overload relay is a device that has three current sensing elements and protects the motor from an over current. Each phase going from the contactor to the motor passes through an overload relay current-sensing element. The overload relay has a selectable current setting based on the full load amp rating of the motor. If the overload current exceeds the setting of the relay for a sufficient length of time, a set of contacts opens to protect the motor from damage.

This article shows how to wire various motors using the FUJI series of contactors sold by Automation Direct. Other brands of contactors may be wired the same or similarly. Consult the manufacturer's wiring diagrams for other brands of contactors.

There are four basic wiring combinations:
- Full-voltage non-reversing 3-phase motors.
- Full-voltage reversing 3-phase motors.
- Single-phase motors.
- Wye-delta open transition 3-phase motors.

You must supply a disconnect switch, proper sized wire, enclosures, terminal blocks and any other devices needed to complete your circuit.

WARNING! Use the instructions supplied for each specific device. Failure to do so may result in electrical shock or damage.

The following components will be used (Fig.2.23).

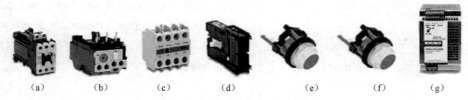

Fig.2.23 Components
(a) Contactors; (b) Overload relays; (c) Aux. contacts; (d) Mechanical interlock;
(e) Normally open pushbutton; (f) Normally closed pushbutton; (g) 24V DC power supply

Full-voltage non-reversing 3-phase motors

The following diagram depicts 3-phase non-reversing motor control with 24V DC control voltage and manual operation (Fig.2.24). We will use a contactor, an auxiliary contact block, an overload relay, a normally open start pushbutton, a normally closed stop pushbutton, and a power supply with a fuse. The start and stop circuits can also be controlled using PLC inputs and outputs.

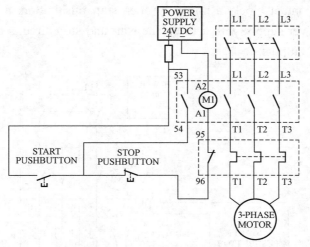

Fig.2.24   Full-voltage non-reversing 3-phase motors

Full-voltage reversing 3-phase motors

This diagram is for 3-phase reversing motor control with 24V DC control voltage (Fig.2.25). It uses two contactors, two auxiliary contact blocks, an overload relay, a mechanical interlock, two normally open start pushbuttons, a normally closed stop pushbutton, and a power supply with a fuse. The forward, reverse, and stop circuits can also be controlled using PLC inputs and outputs.

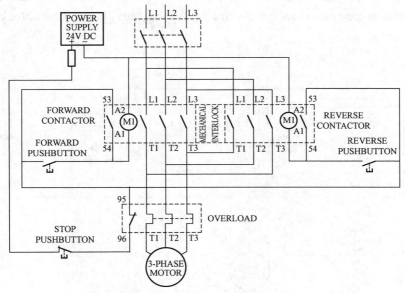

Fig.2.25   Full-voltage reversing 3-phase motors

Full-voltage single-phase motors

This diagram is for single-phase motor control (Fig.2.26). It uses a contactor, an overload relay, one auxiliary contact block, a normally open start pushbutton, a normally closed stop pushbutton, and a power supply with a fuse. The start and stop circuits can also be controlled using PLC inputs and outputs.

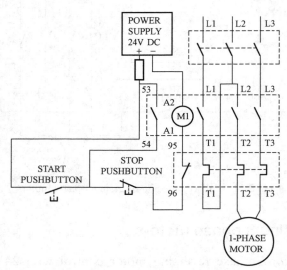

Fig.2.26 Full-voltage single-phase motors

Wye-delta open transition 3-phase motors

The following diagram is shown for 3-phase motor control of a delta-star connection (Fig.2.27). It uses three contactors, an overload relay, one auxiliary contact block, a normally

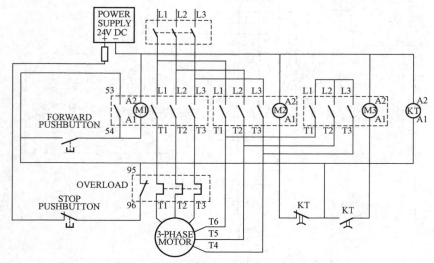

Fig.2.27 Wye-delta open transition 3-phase motors

open start pushbutton, a normally closed stop pushbutton, an on delay timer of 0~20 seconds and a power supply with a fuse. The start, stop, and timing circuits can also be controlled using PLC inputs and outputs.

Technical Words and Expressions

thermal overload		热过载
fuse	n.	熔丝
resistive and lighting load		阻性照明负载
over current		过电流
overload relay		过载继电器
mechanical interlock		机械联动装置，机械联锁装置

Comprehension

1. A motor starter is _____.
   A. used to control the speed of an induction motor
   B. an operator or a controller
   C. controlled by an operator or a controller
   D. a device which only has two components

2. When apply a voltage to the coil of a contactor, the normally open contact will _____ and the normally closed contact will _____.
   A. open, open   B. open, close   C. close, close   D. close, open

3. The overload relay _____.
   A. is not necessary for motor start
   B. can sense voltage
   C. is a current-sensing element
   D. will cut the power supply to the motor in Fig.2.24 when there is over current

4. In Fig.2.25, when reverse contactor is pressed _____.
   A. the motor will run reversely   B. the motor will run forwardly
   C. the contacts of M1 are closed   D. the contacts of M2 are opened

5. In Fig.2.27, when forward pushbutton is pressed _____.
   A. coil of M2 is energized   B. coil of delay timer is energized
   C. motor starts with full voltage.   D. all above

Unit 10 Pulse Width Modulation Motor Control Theory

It is important to be able to control the speed of the robot motors so that you can accurately drive the robot in the direction you want. If you don't have fine enough speed control, the robot is likely to flit left and right and miss your target altogether!

With low powered motors that only consume just a few amps, you could use a variable resistor in series with the motor to vary the current, and therefore vary the speed of the motor[1]. For small motors the resulting power being dissipated in this variable resistor won't be too great and you won't have large heating problems to cope with. This method is often used on small toy models as it is the simplest form of speed control.

When you start to use motors capable of driving heavyweight robots that have stall currents in excess of 100A, a series variable resistor will need to dissipate large amounts of power. Apart from having to be a very low value resistor it is likely to also need extensive heat sinking if you are to stop it glowing red hot[2]! You are also simply wasting energy in heat that would otherwise allow you to keep the robot running for longer periods of time. For these applications, Pulse width modulation (PWM) motor control is a good choice.

PWM duty cycle

A better way of doing speed control is to switch the power to the motor on and off very quickly. The speed of a motor is proportional to the average voltage supplied across it, so if you switch the voltage on and off quickly enough, the motor only "sees" the average voltage. Because of the weight of the motor armature and its inertia, the motor speed won't vary noticeable between each on and off pulse, provided the pulses are short enough and close enough together[3]. To vary the speed of the motor, the duty cycle times of the on and off periods can be varied as shown in Fig.2.28.

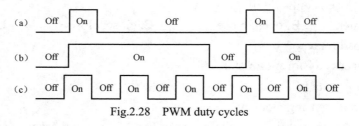

Fig.2.28 PWM duty cycles

On line (a) the on pulses are quite a bit shorter than the off pulses, so the motor will be moving relatively slowly.

To increase the motor speed you need to increase the on period relative to the off period so that the average voltage is increased. On line (b) this was done by making the on pulse longer while at the same time reducing the length of the off period. The time between the start of each successive on pulse remains the same, but the duty cycle has changed.

On line (c), the period of the on pulse remains the same, but the length of the off pulse is reduced. The effect is to increase the average voltage, but this also means there are more pulses per second than in line (b).

If the frequency of the pulses is low enough (less than say 20kHz) you are likely to hear the motors buzz as they run. There is nothing wrong with this, it is just the armature vibrating in sympathy with the pulses. You can observe this effect on modern day trains which buzz noticeably as they are driven, indicating that they too are using PWM as a speed control method. In contrast, the old "slam door" type trains that use resistors to drop the voltage, do not buzz, but are less efficient.

MOSFET devices

The devices normally used to switch the power "on" and "off" are Metal Oxide Semiconductor Field Effect Transistors or MOSFETs. These devices are used because they can switch very large currents under the control of a low signal level voltage.

The normal symbol of a MOSFET is shown below next to a relay type equivalent circuit (Fig.2.29). A purist may not agree that the relay is an exact equivalent circuit, but it does help to show how a MOSFET operates when it is used as a power switching device.

The device switches power between its Drain and Source terminals under control of the signal on the Gate terminal. The threshold at which the MOSFET will switch ON depends on the individual MOSFETs. Any Gate voltage above that "snaps" the MOSFET on faster. However, there is a significant input capacitance on the Gate terminal such that when you want to turn the device OFF again, you will have to discharge this input capacitance, which can slow down the switching time a bit[4]. You therefore have to choose the upper Gate voltage carefully so that you get reasonable ON and OFF switching times.

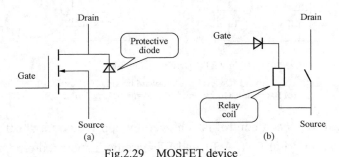

Fig.2.29 MOSFET device
(a) MOSFET device; (b)Equivalent relay circuit

H-Bridge drive configuration

To obtain full control of a motor in both a forwards and backwards direction, an H-bridge configuration is often used as shown in Fig.2.30.

As the name suggests, the MOSFETs are configured around the motor in an H format, such that if VT1 and VT4 are switched on, the motor turns in one way, and if VT2 and VT3 are switched ON, then the motor turns in the other direction. If, for some peculiar reasons you have VT1 and VT3 on together, or VT2 and VT4 on together, you have a smoke making machine as the MOSFET directly short the power rails together and will

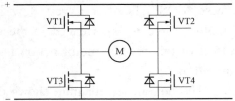

Fig.2.30 Full H-bridge configuration

most likely self destruct in a matter of milliseconds[5]. This situation is normally called "shoot through" and should never be allowed to happen.

Driving ON cycles, and braking OFF cycles

As well as being able to drive a motor using an H-bridge configuration, you can also provide a braking action. The diagram in Fig.2.31 shows how this is achieved.

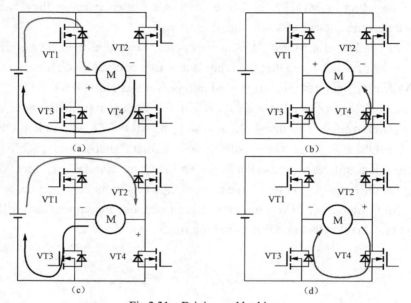

Fig.2.31 Driving and braking
(a) Forward "On"cycle; (b) Forward braking "Off"cycle;
(c) Backward "On"cycle; (d) Backward braking "Off"cycle

As described earlier, speed control is achieved by applying on and off pulses to motor. In the top left configuration, MOSFETs VT1 and VT4 are both switched on during the period and provide power to the motor to make it rotate in a forward direction. During the off period MOSFET VT1 is turned off, and VT3 is turned on in its place, removing the battery from the motor. With the power removed, the motor now acts as a generator and produced an induced voltage across its terminals. The motor now being the source of the voltage means the current flows in the opposite direction through VT3 and VT4, which are both turned on[6]. These MOSFETs offer a very low impedance path for this current, therefore a strong "motor" force is experienced in the opposite direction to that during the period. The net effect of this reverse motor action is a braking force acting on the motor in the backward direction.

The bottom left diagram shows the on cycle situation when the motor is driven in the backwards direction by turning on MOSFETs VT2 and VT3. During the off cycle, VT2 is turned off and VT4 on in its place, and again the motor acts as a generator with the induced currents again flowing through VT3 and VT4 as shown in the bottom right diagram. The

MOSFETs once more offer a low impedance path for these induced currents and the net effect this time is a braking force in the forward direction.

Technical Words and Expressions

Pulse Width Modulation			脉宽调制
dissipate	['disipeit]	v.	消耗，浪费
heavyweight	['heviweit]	n.	重负荷
stall current			堵转电流
average voltage			平均电压
inertia	[i'nə:ʃə]	n.	惯性，惯量
duty cycle			占空因数，占空比
vibrate	[vai'breit]	v.	摆动，振动
in sympathy with			和…一致
buzz	[bʌz]	n.	蜂鸣，嗡嗡声
slam door			摔门式列车
threshold	['θreʃhəuld]	n.	上限，下限，阈值
H-Bridge drive			桥式驱动器
power rail			电源线
millisecond	['mili,sekənd]	n.	毫秒
shoot through			直通
braking force			制动力

Notes

[1] With low powered motors that only consume just a few amps, you could use a variable resistor in series with the motor to vary the current, therefore vary the speed of the motor.

译文：对于仅仅消耗几安培的小功率电动机，可以通过在电动机中串联可变电阻来改变电流，从而改变电动机的速度。

注释："With"引导了一个复合结构在句中作状语，该结构中又含了一个"that"的定语从句修饰"motors"。

[2] Apart from having to be a very low value resistor it is likely to also need extensive heat sinking if you are to stop it glowing red hot!

译文：如果你不想它热得发红，那么除了是一个非常小的电阻外，它还可能要一个散热片。

注释：该句为"if"引导的条件状语从句，表示"如果"，主句中"Apart from…resistor"部分为介词短语作主句的状语，主句为"it is likely to also need extensive heat sinking"。

[3] Because of the weight of the motor armature and its inertia, the motor speed won't vary noticeable between each on and off pulse, provided the pulses are short enough and close enough together.

译文：假定脉冲足够短且距离足够近，那么由于电动机电枢的质量和它的惯性作用，电动机的转速在每个脉冲的通断间将不会明显变化。

注释：该句为"provided"引导假设状语从句。

[4] However, there is a significant input capacitance on the Gate terminal such that when you want to turn the device OFF again, you will have to discharge this input capacitance, which can slow down the switching time a bit.

译文：可是，栅极上有一个明显的输入电容，因此当再次关断元件时，必须给输入电容放电，这将会使关断时间稍微加长。

注释："such that"引导结果状语从句作"因此"解释，该句主句为"there is …terminal"，从句为"that when you …a bit"。而从句部分又是一个"when"引导时间状语从句，该时间状语从句部分又含有一个"which"引导的非限定性定语从句修饰"input capacitance"。

[5] If, for some peculiars reason you have VT1 and VT3 on together, or VT2 and VT4 on together, you have a smoke making machine as the MOSFET directly short the power rails together and will most likely self destruct in a matter of milliseconds.

译文：如果由于某些特殊原因让VT1和VT3同时接通，或VT2和VT4同时接通，那么电源就会被烧坏，因MOSFET直接将电源线短接在一起从而很可能大概在几毫秒内损坏电源。

注释："If"引导条件状语从句，主句为"you have a smoke … milliseconds"。主句中"making machine"为"smoke"的定语直译为"围绕在电机周围的烟雾"，联系上下文该句可意译为"电源就会被烧坏"，"as"作为"出现烟雾"原因。

[6] The motor now being the source of the voltage means the current flows in the opposite direction through VT3 and VT4, which are both turned on.

译文：电动机现在成了一个电压源，这就意味着VT3、VT4被同时接通后，电流通过VT3、VT4以反方向流动。

注释：该句结构是，主语部分"The motor now being the source of the voltage"＋谓语"means"＋宾语从句"the current … and VT4"。"which"引导非限定性定语从句，修饰"VT3 and VT4"。

Exercises

I. Mark the following statements with T (true) or F (false) according to the text.

1. With high powered motors, you could use a variable resistor in series with the motor to vary the speed of the motor. ( )

2. To increase the motor speed you need to increase the on period relative to the off period so that the average current is increased. ( )

3. In contrast, the old "slam door" type trains that use resistors to drop the voltage, do not buzz, but are less efficient. ( )

4. MOSFET switches power between its Gate and Source terminals under control of the signal on the Drain terminal. ( )

5. In H-Bridge driver configuration speed control is achieved by applying on pulses to motor. ( )

II. Complete the following sentences

1. For small motors the resulting power being _____ in variable resistor won't be too great and you won't have large _____ to cope with.

2. A better way of doing speed control is to _____ to the motor on and off very quickly. The speed of a motor is proportional to _____ supplied across it, so if you switch the voltage on and off quickly enough, the motor only "sees" the average voltage.

3. Because of the weight of _____ and _____, the motor speed won't vary noticeable between each on and off _____, provided the pulses are short enough and close enough together.

4. If the _____ of the _____ is low enough (less than say 20kHz) you are likely to hear the motors _____ as they run. it is just _____.

5. To obtain full control of a motor in both _____ and _____ direction, _____ is often used.

Reading 10 Speed and Position Control of DC Motors

Closed loop vs. open loop control

In general open loop control means that you send electrical signals to an actuator to perform a certain action, like connecting a motor to a battery for example. In this scheme of control, there is no any mean for your controller to make sure the task was performed correctly and it often need human intervention to obtain accurate results. A very simple example of open loop control, is the remote controller of an RC toy car; you —— the human —— have to constantly check the position and the velocity of the car to adapt to the situation and move the car to the desired place.

But what if you could let the electronics handle a part, if not all of the tasks performed by a human in an open loop controller, while obtaining more accurate results with extremely short response time? This what is called closed loop control. In order to be able to build a closed loop controller, you need some mean of gaining information about the rotation of the shaft like the number of revolutions executed per second, or even the precise angle of the shaft. This source of information about the shaft of the motor is called "feed-back" because it sends back information from the controlled actuator to the controller.

Fig.2.32 shows clearly the difference between the two control schemes. Both types have a controller that gives orders to a driver, which is a power circuit (usually an H-bridge) that drives the motor in the required direction. It is clear that the closed loop system is more complicated because it needs a "shaft encoder" which is a devise that will translate the rotation of the shaft into electrical signals that can be communicated to the controller.

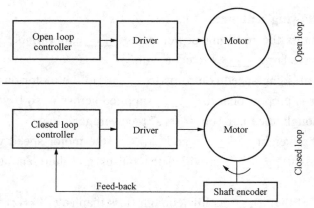

Fig.2.32　Closed loop vs. open loop

In other words, a closed loop controller will regulate the power delivered to the motor to reach the required velocity. If the motor is to turn faster than the required velocity, the controller will deliver less power to the motor. Controlling the electrical power delivered to the motor is usually done by Pulse Width Modulation.

A closed loop controller can be an analog circuit, a digital circuit made of logic gates, or a microcontroller. Generally, a microcontroller is the option that will provide more design flexibility. Recent microcontrollers running at very high clock rates can completely replace similar analog controllers, and can even be cheaper.

In a closed loop system, a microcontroller will have two main tasks:

（1）Constantly adjust the average power delivered to the motor to reach the required velocity.

（2）Precisely calculate the position/angle of the motor's output shaft.

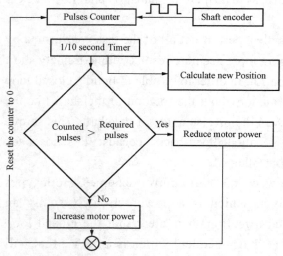

Fig.2.33　Schematic of motor speed control

As you can see in Fig.2.33, the shaft encoder will provide the microcontroller's internal counter with a sequence of pulses that correspond to the rotation of the motor. A timer is set to execute two software routine every 1/10th of a second (which is just an arbitrary value). One of those software routines is to recalculate the actual angle of the shaft or the total number of revolutions.

Then, another software routine is executed to control the speed of the motor by comparing the number of counted pulses with a fixed number which is referred to as the "required pulses". The "required pulses" corresponds to the desired speed, and the "counted pulses" corresponds to the actual speed of the motor.

Chapter 2　Electric Machines and Motor Control

Finally, as you can notice in the schematic, it's all a matter of comparing those two values and constantly adjusting the power delivered to the motor. Note that choosing the right timing between each execution of this routine can dramatically improve overall system stability and performance, especially on low quality motors.

Controlling the power delivered to the motor to control its speed. Recalling the Fig.2.32, a closed loop system contains a controller and a driver. The driver on its own —— which is usually an H-bridge —— cannot control the velocity of the motor. The most common technique to do so is to let the controller turn the driver ON and OFF at very high rates, changing the ratio between the ON and OFF time to control the speed of the motor (Fig.2.33). This is what is called PWM or pulse width modulation.

Technical Words and Expressions

open loop control		开环控制
intervention	n.	干涉
velocity	n.	速度，速率，迅速
response time		响应时间
closed loop control		闭环控制
shaft encoder		转轴编码器

Comprehension

1. General open loop control _____.

　　A. sends electrical signals to a controller to perform a certain action

　　B. has mean for your controller to make sure the task was performed correctly

　　C. often needs human intervention to obtain accurate results

　　D. needs some mean of gaining information about the rotation of the shaft

2. _____ obtains more accurate results with extremely short response time.

　　A. Open loop circuit　　　　　　　B. Closed loop control

　　C. Open loop control　　　　　　　D. Closed loop circuit

3. Which of the following about open loop and closed loop control of motor is not right? _____.

　　A. Both have a controller that gives orders to a driver

　　B. Both have a power circuit which usually is an H bridge that drives the motor in the required direction

　　C. The open loop system is more complicated

　　D. "Shaft encoder" is a devise that will translate the rotation of the shaft into electrical signals

4. A closed loop controller can be _____.

A. an analog circuit B. a digital circuit made of logic gates
C. a microcontroller D. all above

5. The "_____" corresponds to the _____, and the "_____" corresponds to the _____.
   A. required pulse, desired speed, actual speed of the motor, counted pulses
   B. counted pulse, desired speed, required pulses, actual speed of the motor
   C. desired speed, required pulses, counted pulses, actual speed of the motor
   D. required pulse, desired speed, counted pulses, actual speed of the motor

Unit 11 Variable Frequency Drive

A variable frequency drive (VFD in Fig.2.34) is a system for controlling the rotational speed of an alternating current (AC) electric motor by controlling the frequency of the electrical power supplied to the motor. A variable frequency drive is a specific type of adjustable-speed drive. Variable frequency drives are also known as adjustable-frequency drives (AFD), variable speed drives (VSD), AC drives, or inverter drives. Since the voltage is varied along with frequency, these are sometimes also called VVVF (variable voltage variable frequency) drives.

Variable frequency drives are widely used. For example, in ventilations systems for large buildings, variable frequency motors on fans save energy by allowing the volume of air moved to match the system demand. Variable frequency drives are also used on pumps and machine tool drives.

Fig.2.34 Small Variable Frequency Drive (VFD)

Operating principle

The synchronous speed of an AC motor is determined by the frequency of the AC supply and the number of poles in the stator winding, according to the relation:

$$RPM = \frac{120f}{p}$$

Where:
  RPM —— Revolutions per minute;
  f —— AC power frequency (hertz);
  p —— Number of poles (an even number).

The constant 120 is 60 seconds per minute multiplied by 2 poles per pole pair. Sometimes 60 is used as the constant and p is stated as pole pairs rather than poles. By varying the frequency of the voltage applied to the motor, its speed can be changed.

Synchronous motors operate at the synchronous speed determined by the above equation.

Chapter 2 Electric Machines and Motor Control

The speed of an induction motor is slightly less than the synchronous speed.

VFD types

Drives can be classified as: constant voltage, constant current, cycloconverter.

In a constant voltage converter, the intermediate DC link voltage remains approximately constant during each output cycle. In constant current drives, a large inductor is placed between the input rectifier and the output bridge, so the current delivered is nearly constant. A cycloconverter has no input rectifier or DC link and instead connects each output terminal to the appropriate input phase.

The most common type of packaged VF drive is the constant-voltage type, using pulse width modulation to control both the frequency and effective voltage applied to the motor load[1].

VFD structure

All VFDs have the same structure, which includes a rectifier, filter and inverter, as shown in Fig.2.35. The rectifier converts 3~0 AC line power to DC power. The components used in the rectifier are typically thyristors and/or diodes. The filter sits between the rectifier & inverter and provides harmonic and power ripple filtration using inductors and capacitors[2]. These components work to respectively smoothen and regulate the current and voltage supplied to the motor. The inverter portion typically consists of thyristors or IGBTs that are carefully controlled to sequence the proper voltage and current to the phase windings of the motor, depending on the speed and load required[3]. In case of IGBTs, control circuit is connected to the gate, but this circuitry is less complex and does not require polarity reversal.

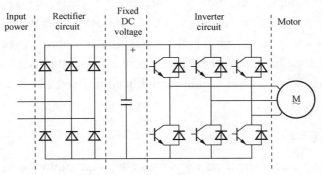

Fig.2.35 PWM AC variable speed drive

Inverters provide the user with tremendous flexibility by controlling two main elements of a three-phase induction motor, i.e., torque and speed. Motor speed and many operating parameters may be infinitely adjusted. Some of the other types of inverters include current source inverters, voltage source inverters; load commuted inverters, PWM inverters, vector control, etc. Each type of inverter has its own sets of advantages and disadvantages and should be selected according to particular requirements. The present economics favours PWM drive for applications under 200HP.

VFD operator interface

Fig.2.36 VFD operator interface

The operator interface provides a means for an operator to start and stop the motor and adjust the operating speed. Additional operator control functions might include reversing and switching between manual speed adjustment and automatic control from an external process control signal[4]. The operator interface often includes an alphanumeric display and/or indication lights and meters to provide information about the operation of the drive. An operator interface keypad and display unit is often provided on the front of the VFD controller as shown in Fig.2.36. The keypad display can often be cable-connected and mounted a short distance from the VFD controller. Most are also provided with input and output (I/O) terminals for connecting pushbuttons, switches and other operator interface devices or control signals. A serial communications port is also often available to allow the VFD to be configured, adjusted, monitored and controlled using a computer.

VFD advantages

VFDs eliminate the need for energy wasting throttling mechanisms like control valves and outlet dampers. These are mainly installed on loads where torque increases with speed including centrifugal pumps, fans, blowers, compressors, etc.[5]. VFDs are used to improve process control, which in turn increases the efficiency of an operating system. These drives reduce mechanical stress by employing soft starters. It also improves the power factor of the whole electrical system. The other benefits include the reduction and/or elimination of motor starters, less stress on the AC motor windings and bearings, and a decrease in stress and wear on the pump or fan.

VFD disadvantages

The non-pure sine wave output power causes additional motor heating primarily because of harmonics. Depending upon the manufacturer and the type of variable frequency controller, the input voltage and current waveform to the motor will be distorted by varying degrees. Harmonic currents generate no useful torque at the motor shaft. These create additional losses.

However, VFDs can be reliable and trouble-free if they are selected and installed properly. VFDs can cause damage to the motors they drive and cause a wide range of problems in the associated electrical system if they are not selected properly. Most of the problems arise from distortion of the voltage and current waveforms in the input and output of the drive, commonly called harmonic distortion.

Chapter 2 Electric Machines and Motor Control

Technical Words and Expressions

variable-frequency drive			变频器
rotational speed			转动速度，角速度
adjustable-speed drive			调速器
adjustable-frequency drive			调频器
variable-speed drive (VSD)			变速器
VVVF(variable voltage variable frequency)dvive			变压变频器
ventilation	[venti'leiʃən]	n.	通风，流通空气
synchronous speed			同步速度，同步转速
stator winding			定子绕组
even number			偶数
constant voltage			恒压
constant current			恒流
cycloconverter	[ˌsaikləukən'və:tə]	n.	循环换流器，双向离子变频器
filtration	[fil'treiʃən]	n.	过滤
inverter	[in'və:tə]	n.	逆变器
power ripple			电源纹波，电源脉动
smoothen	['smu:ðən]	v.	使平滑，使平和
polarity reversal			极性变换
operating parameter			运行参数
load commuted inverter			负载换流逆变器
vector control			矢量控制
operator interface			操作界面
manual speed adjustment			手动调速
alphanumeric display			字母数字显示器
indication light			指示灯
meter	['mi:tə]	n.	仪表
pushbutton		n.	按钮
throttling mechanism			节流装置
control valve			控制阀，调节阀
damper	['dæmpə]	n.	风门；节气阀
centrifugal pump			离心泵，离心抽机
soft starter			软启动器
power factor			功率因数
non-pure sine wave			非正弦波
harmonic current			谐波（正弦）电流

trouble-free 无故障
harmonic distortion 谐波畸变（失真）

 Notes

[1] The most common type of packaged VF drive is the constant-voltage type, using pulse width modulation to control both the frequency and effective voltage applied to the motor load.

译文：最常见的封装式变频器是恒压型，它采用脉宽调制技术来控制频率及加在电动机负载上的有效电压。

注释：该句中"using…"部分为现在分词作方式状语，"applied to the motor load"部分则为"effective voltage"的定语。

[2] The filter sits between the rectifier & inverter and provides harmonic and power ripple filtration using inductors and capacitors.

译文：滤波器位于整流器和逆变器之间，通过使用电感和电容来滤去谐波和电源纹波（电源脉动）。

注释：第一个"and"连接了两个相同主语的并列句，第二个"and"连接了"provides"的两个宾语。而："using inductors and capacitors"为现在分词短语作条件状语。

[3] The inverter portion typically consists of thyristors or IGBTs that are carefully controlled to sequence the proper voltage and current to the phase windings of the motor, depending on the speed and load required.

译文：逆变部分通常由晶闸管或IGBT组成，它将根据转速和负载的要求来恰当地控制晶闸管或IGBT，从而给电动机各相绕组提供合适的电压和电流。

注释：该句为"that"引导的定语从句修饰"thyristors or IGBTs"，但翻译时由于从句部分过长，因而把从句部分单独作为一句翻译。而从句中的"depending on…"部分为分词作独立成分。

[4] Additional operator control functions might include reversing and switching between manual speed adjustment and automatic control from an external process control signal.

译文：此外它的操作控制功能还包括反转以及手动和自动调速间的切换，自动调速是由外部控制信号来控制的。

注释：该句翻译时把"from an external process control signal"部分单独作一句翻译，否则该部分放置在主句中翻译，会使主句显得冗长。

[5] These are mainly installed on loads where torque increases with speed including centrifugal pumps, fans, blowers, compressors, etc.

译文：它们主要安装在转矩随转速变化而变化的负载上，如离心泵、风扇、吹风机、压缩机等。

注释：该句为"where"引导的定语从句修饰"loads"。

Chapter 2 Electric Machines and Motor Control

Exercises

I. Mark the following statements with T (true) or F (false) according to the text.

1. A variable-frequency drive (VFD) is a system for controlling the rotational speed of a direct current (DC) electric motor. ( )

2. Induction motors operate at the synchronous speed determined by the above equation. The speed of Synchronous motor is slightly less than the synchronous speed. ( )

3. In a constant voltage converter, the intermediate DC link voltage remains approximately constant during each output cycle. ( )

4. The pure sine wave output power causes additional motor heating primarily because of harmonics. ( )

5. The operator interface provides a means for an operator to start and stop the motor and adjust the operating speed. ( )

II. Complete the following sentences.

1. A variable frequency drive is a specific type of _____. Variable-frequency drives are also known as _____, variable-speed drives (VSD), AC drives, or _____.

2. _____ of an AC motor is determined by the frequency of the AC supply and the number of _____ in the _____.

3. The most common type of packaged VF drive is _____ type, using _____ to control both the frequency and _____ applied to the motor load.

4. All VFDs have the same structure, which includes _____, filter and _____. The filter sits between the _____ and provides _____ filtration using _____ and capacitors.

5. VFDs are used to improve _____, which in turn increases the _____ of an operating system. These drives reduce mechanical stress by employing _____. It also improves _____ of the whole electrical system.

Reading 11 Vector Control (Motor)

Vector control (also called Field Oriented Control, FOC) is one method used in variable frequency drives to control the torque (and thus finally the speed) of three-phase AC electric motors by controlling the current fed to the machine.

The Stator phase currents are measured and converted into a corresponding complex (space) vector. This current vector is then transformed to a coordinate system rotating with the rotor of the machine. For this the rotor position has to be known. Thus at least speed measurement is required. The position can then be obtained by integrating the speed.

Then the rotor flux linkage vector is estimated by multiplying the stator current vector with magnetizing inductance L_m and low-pass filtering the result with the rotor no-load time constant L_r/R_r, that is the ratio of the rotor inductance to rotor resistance.

Using this rotor flux linkage vector the stator current vector is further transformed into a coordinate system where the real x-axis is aligned with the rotor flux linkage vector.

Now the real x-axis component of the stator current vector in this rotor flux oriented coordinate system can be used to control the rotor flux linkage and the imaginary y-axis component can be used to control the motor torque.

Typically PI-controllers are used to control these currents to their reference values. However, bang-bang type current control, that gives better dynamics, is also possible.

With PI-controllers the outputs of the controllers are the x-y components of the voltage reference vector for the stator. Usually due to the cross coupling between the x-axis and y-axis a decoupling term is further added to the controller output in order to improve the control performance when big and rapid changes in speed, current and flux linkage take place. Usually the PI-controller also needs low-pass filtering of either the input or output of the controller in order to prevent the current ripple due to transistor switching from being amplified excessively and unstabilizing the control. Unfortunately, the filtering also limits the dynamics of the control system. Thus quite high switching frequency (typically more than 10kHz) is required in order to allow only minimum filtering for high performance drives such as servo drives.

Next the voltage references are first transformed to the stationary coordinate system (usually through rotor d-q coordinates) and then fed into a modulator that using one of the many Pulse Width Modulation (PWM) algorithms defines the required pulse widths of the stator phase voltages and controls the transistors (usually IGBTs) of the inverter according to these.

This control method implies the following properties of the control (Fig.2.37):

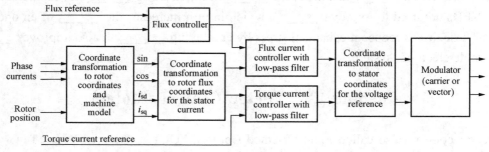

Fig.2.37 Vector control of motor

Speed or position measurement or some sort of estimation is needed;

Torque and flux can be changed reasonably fast, in less than 5~10 milliseconds, by changing the references;

The step response has some overshoot if PI control is used;

The switching frequency of the transistors is usually constant and set by the modulator;

The accuracy of the torque depends on the accuracy of the motor parameters used in the control. Thus large errors due to for example rotor temperature changes often are encountered.

Reasonable processor performance is required, typically the control algorithm has to be calculated at least every millisecond.

Although the vector control algorithm is more complicated than the Direct Torque Control (DTC), the algorithm is not needed to be calculated as frequently as the DTC algorithm. Also the current sensors need not be the best in the market. Thus the cost of the processor and other control hardware is lower making it suitable for applications where the ultimate performance of DTC is not required.

Technical Words and Expressions

vector control		矢量控制
Field Oriented Control		磁场定向控制
coordinate system		坐标系
magnetizing inductance		磁化电感
multiply	v.	乘
low-pass filtering		低通滤波
orient	vt.	使适应，确定方向
reference value		参考值
bang-bang type current control		继电器式电流控制
overshoot	n.	超调
Direct Torque Control		直接转矩控制
algorithm	n.	运算法则，算法

Comprehension

1. Vector control is _____.
   A. also called Field Oriented Control, FOC
   B. used to control the speed of three-phase AC electric motors
   C. used to control the torque of three-phase AC electric motors by controlling the input current
   D. all above

2. Which of the following is right? _____.
   A. To convert stator phase currents into a coordinate system rotating with the rotor of the motor, you have to know the rotor position
   B. Speed measurement can be used to obtain the rotor position by integrating the speed
   C. The rotor flux linkage vector is estimated by multiplying the stator current vector with magnetizing inductance L_m and low-pass filtering the result with the rotor no-load time the ratio of the rotor inductance to rotor resistance
   D. all above

3. In the process of Vector control, _____.
   A. PI-controllers are used to control voltage to their reference values
   B. the outputs of the PI controllers are the x-y components of the voltage reference vector for the stator
   C. usually the PI-controller also needs high-pass filtering of either the input or output of the controller in order to prevent the current ripple
   D. modulator uses PWM algorithms to define the required pulse widths of the rotor phase voltages
4. Vector control method has the properties that _____.
   A. speed or position measurement or some sort of estimation is not needed
   B. the step response has some overshoot if P control is used
   C. the switching frequency of the transistors is usually variable
   D. rotor temperature changes often result in the inaccuracy of the torque
5. Compared to Direct Torque Control, the vector control algorithm _____.
   A. is more complex
   B. needs to be calculated frequently
   C. needs the best current sensors in the market
   D. costs more

Chapter 3
Industry Computer Control Technology

Unit 12 Overview of Programmable Logic Controller (PLC)

What is a PLC?

Programmable logic controllers, also called programmable controllers or PLCs, are solid-state members of the computer family, using integrated circuits instead of electromechanical devices to implement control functions[1]. They are capable of storing instructions, such as sequencing, timing, counting, arithmetic, data manipulation, and communication, to control industrial machines and processes.

Programmable controllers have many definitions. However, PLCs can be thought of in simple terms as industrial computers with specially designed architecture. It is designed to read digital and analog inputs from various sensors, execute a user defined logic program, and write the resulting digital and analog output values to control industrial machines and processes (Fig.3.1).

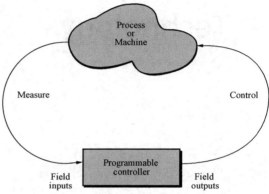

Now PLCs are used in many different industries and machines such as packaging and semiconductor machines. Well known PLC brands are Siemens, Allen-Bradley, ABB, Mitsubishi, Omron, and General Electric.

Fig.3.1 The principle of PLC

Components of the PLC

A schematic diagram of a programmable logic controller is presented in Fig.3.2. The basic components of the PLC are the following.

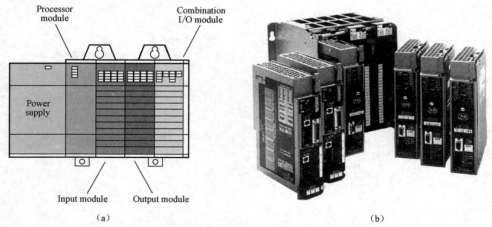

Fig.3.2 Diagram of the programmable logic controller

Processor

The processor (sometimes called a CPU), as in the self contained units, is generally specified according to memory required for the program to be implemented[2]. In the modularized versions, capability can also be a factor. This includes features such as higher math functions, PID control loops and optional programming commands. The processor consists of the microprocessor, system memory, and serial communication ports for printer, PLC LAN link and external programming device and, in some cases, the system power supply to power the processor and I/O modules.

Mounting rack

This is usually a metal framework with a printed circuit board backplane which provides means for mounting the PLC input/output (I/O) modules and processor. Mounting racks are specified according to the number of modules required to implement the system. The mounting rack provides data and power connections to the processor and modules via the backplane. For CPUs that do not contain a power supply, the rack also holds the modular power supply. There are systems in which the processor is mounted separately and connected by cable to the rack. The mounting rack can be available to mount directly to a panel or can be installed in a standard "19" wide equipment cabinet. Mounting racks are cascadable so several may be interconnected to allow a system to accommodate a large number of I/O modules.

Input and output modules

Input/Output units are the interfaces between the internal PLC systems and the external processes to be monitored and controlled. Since the PLC is a logic based device with a typical operating voltage of 5 volts and the external processes usually demand higher powers and currents, the I/O modules are optically isolated[3]. The typical I/O operating voltages are 5~240V DC (or AC) and currents from 0.1A up to several amperes. The I/O modules are designed in this way to minimize or eliminate the need for any intermediate circuitry between the PLC and the process to be controlled. Small PLC units would have around 40 I/O connections, while larger ones having more than 128 with either local or remote connections and extensive upgrade capabilities.

Power supply

The power supply specified depends upon the manufacturer's PLC being utilized in the application. As stated above, in some cases a power supply capable of delivering all required power for the system is furnished as part of the processor module[4]. If the power supply is a separate module, it must be capable of delivering a current greater than the sum of all the currents needed by the other modules. For systems with the power supply inside the CPU module, there may be some modules in the system which require excessive power not available from the processor either because of voltage or current requirements that can only be achieved through the addition of a second power source[5]. This is generally true if analog or external communication modules are present since these require ±DC supplies which, in the case of analog modules, must be well regulated.

Programming unit

The programming unit allows the engineer or technician to enter and edit the program to be executed. In its simplest form, it can be a hand held device with a keypad for program entry and a display device (LED or LCD) for viewing program steps or functions, as shown in Fig.3.3. More advanced systems employ a separate personal computer which allows the programmer to write, view, edit and download the program to the PLC. This is accomplished with proprietary software available from the PLC manufacturer. This software also allows the programmer or engineer to monitor the PLC as it is running the program. With this monitoring system, such things as internal coils, registers, timers and other items not visible externally can be monitored to determine proper operation. Also, internal register data can be altered if required to fine tune program operation. This can be advantageous when debugging the program. Communication with the programmable controller with this system is via a cable connected to a special programming port on the controller. Connection to the personal computer can be through a serial port or from a dedicated card installed in the computer.

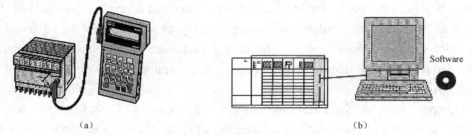

Fig.3.3　Programming unit

(a) The simplest programming unit; (b) More advanced programming unit

Programming languages for PLC

Early PLCs were designed to replace relay logic systems. These PLCs were programmed in "ladder logic", which strongly resembles a schematic diagram of relay logic. Modern PLCs can be programmed in a variety of ways, from ladder logic to more traditional programming languages such as BASIC and C. Another method is State Logic, a Very High Level Programming Language designed to program PLCs based on State Transition Diagrams.

IEC 61131-3 currently defines five programming languages for programmable control systems: FBD (Function block diagram), LD (Ladder diagram), ST (Structured text, similar to the Pascal programming language), IL (Instruction list, similar to assembly language) and SFC (Sequential function chart).

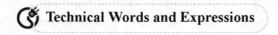

Technical Words and Expressions

programmable logic controller　　　　　　　　　　　　　　　可编程控制器

Chapter 3 Industry Computer Control Technology

electromechanical	[i,lektrəumiˆkænikəl]	adj.	[机]电动机械的，机电的，电机的
sequencing	['siːkwənsiŋ]	n.	先后顺序
timing	['taimiŋ]	n.	定时
counting	['kauntiŋ]	n.	计算
data manipulation			数据操作
architecture	['ɑːkitektʃə]	n.	建筑，结构
brand	[brænd]	n.	商标，牌子
Siemens		n.	[德]西门子
Allen-Bradley		n.	[美]艾伦-布拉德利
Mitsubishi	[mi'tsubiʃi]	n.	[日]三菱
Omron		n.	[日]欧姆龙
processor	['prəusesə]	n.	处理机，处理器
serial communication port			串行通信口
module	['mɔdjuːl]	n.	组合部件，模块
mounting rack			机架
backplane			底板
hand held device			手持式设备
keypad	['kiːpæd]	n.	键盘
download		vt.	下载
register	['redʒistə]	n.	寄存器
debugging	[diː'bʌgiŋ]	n.	调试
ladder logic			梯形图
Very High Level Programming Language			超高级编程语言
State Transition Diagram			状态转换图
FBD (Function block diagram)			功能块图
LD (Ladder diagram)			梯形图
ST (Structured text)			结构文本
IL (Instruction list)			指令表
SFC (Sequential function chart)			顺序功能图

 Notes

[1] Programmable logic controllers, also called programmable controllers or PLCs, are solid-state members of the computer family, using integrated circuits instead of electromechanical devices to implement control functions.

译文：可编程逻辑控制器，又称可编程控制器或 PLC，是计算机家族的一员，它是用集成电路来执行控制任务，而不是机电设备。

注释："also called programmable controllers or PLCs"为过去分词作定语修饰

"Programmable logic controllers","using…"部分为现在分词作状语修饰。

[2] The processor (sometimes called a CPU), as in the self contained units, is generally specified according to memory required for the program to be implemented.

译文：对于整体式 PLC，处理器（有时又叫 CPU），整体式 PLC，通常据执行程序所需的存储器的大小来指定。

注释："required for the program"为过去分词作定语修饰"memory"，"to be implemented"为不定式作定语，修饰"program"。

[3] Since the PLC is a logic based device with a typical operating voltage of 5 volts and the external processes usually demand higher powers and currents, the I/O modules are optically isolated.

译文：由于 PLC 是一个基于逻辑运算的设备，工作电压一般为 5V，而外部生产过程通常需要更高的电压和电流，因此输入/输出模块通常是光隔离的。

注释："Since"引导了一个原因状语从句，从句部分为"Since the PLC is…and currents"，主句部分为"the I/O modules are optically isolated"。从句为"and"连接的两个并列句，其中"with"表示"带有"的意思。

[4] As stated above, in some cases a power supply capable of delivering all required power for the system is furnished as part of the processor module.

译文：如上所说，某些 PLC 的电源是作为处理器模块的一部分，它能够给系统提供各种所需电源。

注释："be furnished as"作为……提供，词句虽然是被动形式，但需按主动形式来翻译。

[5] For systems with the power supply inside the CPU module, there may be some modules in the system which require excessive power not available from the processor either because of voltage or current requirements that can only be achieved through the addition of a second power source.

译文：对于电源内置在 CPU 模块中的 PLC，系统中可能存在一些模块，因为所需的电压或电流只能从附加的第二个电源上得到，这些模块所需的额外电源不能够从处理器上获得。

注释：Which 引导定语从句修饰"modules"，该从句中"(which is) not available from the processor"为修饰"excessive power"的定语。

Exercises

I. Mark the following statements with T (true) or F (false) according to the text.

1. PLCs use integrated circuits to implement control functions.　　　　　　　(　　)

2. The processor (sometimes call a CPU), is generally specified according to memory required for the program to be implemented.　　　　　　　(　　)

3. Mounting rack is usually a metal framework with a printed circuit board backplane which provides means for mounting the PLC input/output (I/O) modules only.　　(　　)

4. The typical I/O operating voltages are 5~24V DC (or AC) and currents from 0.1A up to several amperes. ( )

5. If the power supply is a separate module, it must be capable of delivering a current greater than the sum of all the currents needed by the other modules. ( )

II. Complete the following sentences.

1. It is designed to read _____ from various _____, execute a user defined _____, and write the resulting digital and analog output values to control industrial machines and processes.

2. This includes features such as _____, _____ and _____.

3. The processor consists of the _____, system memory, _____ for printer, PLC LAN link and _____ and, in some cases, the system power supply to power the processor and _____.

4. More advanced systems of PLC employ a separate personal computer which allows the programmer to _____, _____, _____ and _____ the program to the PLC. This is accomplished with _____ available from the PLC _____.

5. Early PLCs were designed to replace _____. These PLCs were programmed in "_____", which strongly resembles a schematic diagram of relay logic. Modern PLCs can be programmed in a variety of ways, from ladder logic to _____ languages such as BASIC and C. Another method is _____.

Reading 12 Process Control System with PLC

Introduction

Generally speaking, process control system is made up of a group of electronic devices and equipment that provide stability, accuracy and eliminate harmful transition statuses in production processes. Operating system can have different form and implementation, from energy supply units to machines. As a result of fast progress in technology, many complex operational tasks have been solved by connecting programmable logic controllers and possibly a central computer. Besides connections with instruments like operating panels, motors, sensors, switches, valves and such, possibilities for communication among instruments are so great that they allow high level of exploitation and process coordination, as well as greater flexibility in realizing a process control system. Each component of a process control system plays an important role, regardless of its size. For example, without a sensor, PLC wouldn't know what exactly goes on in the process. In an automated system, PLC controller is usually the central part of a process control system. With execution of a program stored in program memory, PLC continuously monitors status of the system through signals from input devices. Based on the logic implemented in the program, PLC determines which actions need to be executed with output instruments. To run more complex processes, we can connect more

PLC controllers to a central computer. A real system could look like the one pictured in Fig.3.4.

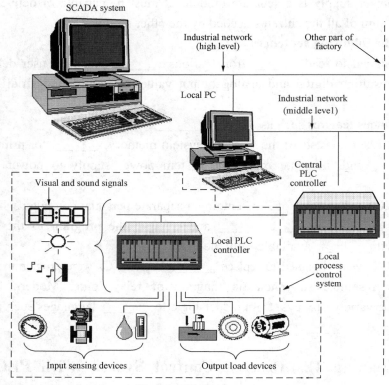

Fig.3.4　SCADA system

Conventional control panel

At the outset of industrial revolution, especially during sixties and seventies, relays were used to operate automated machines, and these were interconnected using wires inside the control panel. In some cases, a control panel covered an entire wall. To discover an error in the system much time was needed especially with more complex process control systems. On top of everything, the lifetime of relay contacts was limited, so some relays had to be replaced. If replacement was required, machine had to be stopped and production too. Also, it could happen that there was not enough room for necessary changes. Control panel was used only for one particular process, and it wasn't easy to adapt to the requirements of a new system. As far as maintenance, electricians had to be very skillful in finding errors. In short, conventional control panels proved to be very inflexible. Typical example of a conventional control panel is given in Fig.3.5.

In this photo you can notice a large number of electrical wires, time relays, timers and other elements of automation typical for that period. Pictured control panel is not one of the more "complicated" ones, so you can imagine what complex ones looked like.

Chapter 3 Industry Computer Control Technology

(a)　　　　　　　　　　　　(b)

Fig.3.5   Control panel with a PLC controller and conventional control panel
(a) Control panel with PLC; (b) Conventional control

Most frequently mentioned disadvantages of a typical control panel are:
- Too much work required in connecting wires;
- Difficulty with changes or replacements;
- Difficulty in finding errors; requiring skillful work force;
- When a problem occurs, hold-up time is indefinite, usually long.

Control panel with a PLC controller

With the invention of programmable controllers, much has changed in how a process control system is designed. Many advantages appeared. Typical example of control panel with a PLC controller is given in the following picture.

Advantages of control panel that is based on a PLC controller can be presented in a few basic points:
- Compared to a conventional process control system, the number of wires needed for connections is reduced by 80%.
- Consumption is greatly reduced because a PLC consumes less than a bunch of relays.
- Diagnostic functions of a PLC controller allow for fast and easy error detection.
- Change in operating sequence or application of a PLC controller to a different operating process can easily be accomplished by replacing a program through a console or using a PC software (not requiring changes in wiring, unless addition of some input or output device is required).
- Needs fewer spare parts.
- It is much cheaper compared to a conventional system, especially in cases where a large number of I/O instruments are needed and when operational functions are complex.
- Reliability of a PLC is greater than that of an electro-mechanical relay or a timer.

Technical Words and Expressions

process control system　　　　　　　　过程控制系统
operating panel　　　　　　　　　　　　控制盘，控制面板
at the outset of　　　　　　　　　　　　最初，开始
diagnostic function　　　　　　　　　　诊断功能

Comprehension

1. In Process control system with PLC, the actions of output devices are determined by _____.
   A. the input signals of PLC
   B. the program in PLC
   C. the logic implemented in the computer
   D. A and B

2. which of the following is right? _____.
   A. At the beginner of industrial revolution, relays were used in conventional control panel
   B. The size of conventional control panel is very small
   C. Conventional control panel is easy to maintain
   D. Conventional control panel can be used for different process control system

3. The most disadvantage of a conventional control panel is _____.
   A. too much work required in connecting wires
   B. difficult in finding errors
   C. lifetime of relay contacts is limited
   D. all above

4. Compare to conventional control panel, control panel with a PLC controller _____.
   A. needs more wires
   B. maintenance is more complicated
   C. can be used to only one operating process easily
   D. is greater reliability

5. A PLC is much cheaper compared to an electro-mechanical relay or a timer, _____.
   A. when control functions are more complex
   B. when a control system needs a large number of input devices
   C. when a control system needs a large number of output devices
   D. all above

Chapter 3 Industry Computer Control Technology

Unit 13 Applications of PLC

PLCs are industrial computers and able to interface with real-world devices such as switches, solenoids and so on. The actual logic of the control system is established inside the PLC by means of a computer program. This program dictates which output gets energized under which input conditions. Although the program itself appears to be a ladder logic diagram, with switch and relay symbols, there are no actual switch contacts or relay coils operating inside the PLC to create the logical relationships between input and output[1]. These are imaginary contacts and coils, if you will. The program is entered and viewed via a personal computer connected to the PLC's programming port.

Consider the circuit shown in Fig.3.6 and PLC program.

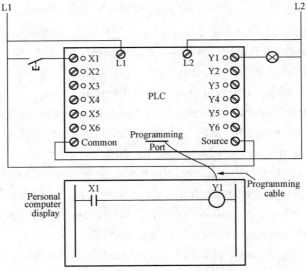

Fig.3.6 A simple PLC circuit

When the pushbutton switch is unactuated (unpressed), no power is sent to the X1 input of the PLC. Following the program, which shows a normally-open X1 contact in series with a Y1 coil, no "power" will be sent to the Y1 coil. Thus, the PLC's Y1 output remains de-energized, and the indicator lamp connected to it remains dark.

If the pushbutton switch is pressed, however, power will be sent to the PLC's X1 input. Any and all X1 contacts appearing in the program will assume the actuated (non-normal) state, as though they were relay contacts actuated by the energizing of a relay coil named "X1"[2]. In this case, energizing the X1 input will cause the normally-open X1 contact will "close," sending "power" to the Y1 coil. When the Y1 coil of the program "energizes", the real Y1 output will become energized, lighting up the lamp connected to it (Fig.3.7).

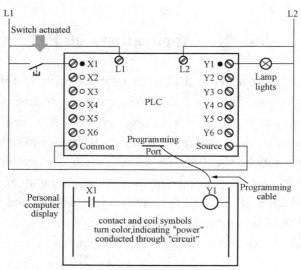

Fig.3.7 PLC circuit with one switch being actuated

It must be understood that the X1 contact, Y1 coil, connecting wires, and "power" appearing in the personal computer's display are all virtual. They do not exist as real electrical components. They exist as commands in a computer program —— a piece of software only —— that just happens to resemble a real relay schematic diagram.

Equally important to understand is that the personal computer used to display and edit the PLC's program is not necessary for the PLC's continued operation[3]. Once a program has been loaded to the PLC from the personal computer, the personal computer may be unplugged from the PLC, and the PLC will continue to follow the programmed commands. I include the personal computer display in these illustrations for your sake only, in aiding to understand the relationship between real-life conditions (switch closure and lamp status) and the program's status ("power" through virtual contacts and virtual coils)[4].

The true power and versatility of a PLC is revealed when we want to alter the behavior of a control system. Since the PLC is a programmable device, we can alter its behavior by changing the commands we give it, without having to reconfigure the electrical components connected to it.

One of the advantages of implementing logical control in software rather than in hardware is that input signals can be re-used as many times in the program as is necessary. For example, take the following circuit and program, designed to energize the lamp if at least two of the three pushbutton switches are simultaneously actuated (Fig.3.8).

To build an equivalent circuit using electromechanical relays, three relays with two normally-open contacts each would have to be used, to provide two contacts per input switch. Using a PLC, however, we can program as many contacts as we wish for each "X" input without adding additional hardware, since each input and each output is nothing more than a single bit in the PLC's digital memory (either 0 or 1), and can be recalled as many times as necessary[5].

Furthermore, since each output in the PLC is nothing more than a bit in its memory as well, we can assign contacts in a PLC program "actuated" by an output (Y) status. Take for instance this next system, a motor start-stop control circuit (Fig.3.9).

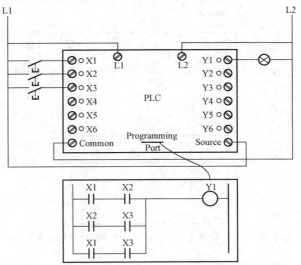

Fig.3.8 PLC circuit with at least two of the three pushbutton switches simultaneously being actuated

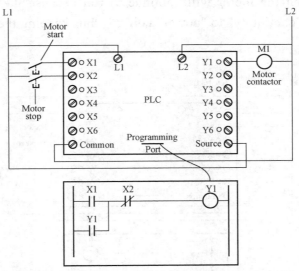

Fig.3.9 PLC circuit with motor start-stop control

The pushbutton switch connected to input X1 serves as the "Start" switch, while the switch connected to input X2 serves as the "Stop." Another contact in the program, named Y1, uses the output coil status as a seal-in contact, directly, so that the motor contactor will continue to be energized after the "Start" pushbutton switch is released. You can see the normally-closed contact X2 appear in a colored block, showing that it is in a closed ("electrically conducting") state.

If we were to press the "Start" button, input X1 would energize, thus "closing" the X1 contact in the program, sending "power" to the Y1 "coil," energizing the Y1 output and applying 120 volt AC power to the real motor contactor coil. The parallel Y1 contact will also "close," thus latching the "circuit" in an energized state (Fig.3.10).

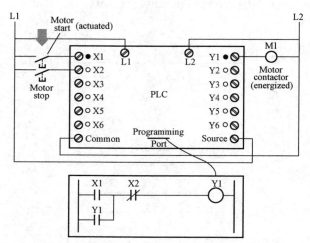

Fig.3.10   PLC circuit with motor start button being actuated

Now, if we release the "Start" pushbutton, the normally-open X1 "contact" will return to its "open" state, but the motor will continue to run because the Y1 seal-in "contact" continues to provide "continuity" to "power" coil Y1, thus keeping the Y1 output energized (Fig.3.11).

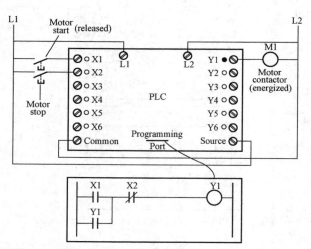

Fig.3.11   PLC circuit with motor start button being released

To stop the motor, we must momentarily press the "Stop" pushbutton, which will energize the X2 input and "open" the normally-closed "contact," breaking continuity to the Y1 "coil" (Fig.3.12).

When the "Stop" pushbutton is released, input X2 will de-energize, returning "contact" X2 to its normal, "closed" state. The motor, however, will not start again until the "Start" pushbutton is actuated, because the "seal-in" of Y1 has been lost (Fig.3.13).

Chapter 3 Industry Computer Control Technology

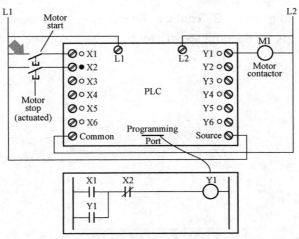

Fig.3.12 PLC circuit with motor stop button being actuated

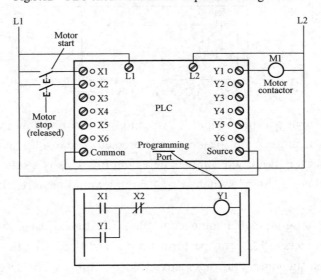

Fig.3.13 PLC circuit and PLC program

An important point to make here is that fail-safe design is just as important in PLC-controlled systems as it is in electromechanical relay-controlled systems. One should always consider the effects of failed (open) wiring on the device or devices being controlled. In this motor control circuit example, we have a problem: if the input wiring for X2 (the "Stop" switch) were to fail open, there would be no way to stop the motor!

The solution to this problem is a reversal of logic between the X2 "contact" inside the PLC program and the actual "Stop" pushbutton switch (Fig.3.14).

When the normally-closed "Stop" pushbutton switch is unactuated (not pressed), the PLC's X2 input will be energized, thus "closing" the X2 "contact" inside the program. This allows the motor to be started when input X1 is energized, and allows it to continue to run when the "Start" pushbutton is no longer pressed. When the "Stop" pushbutton is actuated, input X2 will de-energize, thus "opening" the X2 "contact" inside the PLC program and shutting off the motor. So, we see there is no operational difference between this new design and the previous design.

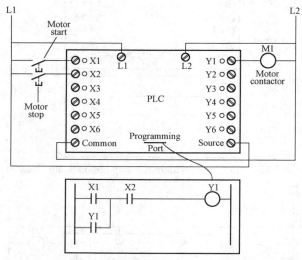

Fig.3.14　PLC circuit and PLC program

However, if the input wiring on input X2 were to fail open, X2 input would de-energize in the same manner as when the "Stop" pushbutton is pressed. The result, then, for a wiring failure on the X2 input is that the motor will immediately shut off. This is a safer design than the one previously shown, where a "Stop" switch wiring failure would have resulted in an inability to turn off the motor.

This section on programmable logic controllers illustrates just a small sample of their capabilities. As computers, PLCs can perform timing functions (for the equivalent of time-delay relays), drum sequencing, and other advanced functions with far greater accuracy and reliability than what is possible using electromechanical logic devices.

Technical Words and Expressions

solenoid	['səulinɔid]	n.	螺线管
dictate	[dik'teit]	n.	指令，指示，命令，规定
energize	['enədʒaiz]	vt.	使…得电
programming port			编程端口
normally-open contact			常开触点（动合触点）
indicator lamp			指示灯
non-normal			非正规的，非常态的
relay schematic diagram			继电器原理图
for…sake			为了…的缘故

Chapter 3　Industry Computer Control Technology

reconfigure	[,riːkən'figə(r)]	v.	重新装配，改装
electromechanical relay			电磁继电器
digital memory			数字存储器
fail-safe design			可靠性设计
time-delay relay			延时继电器

 Notes

[1] Although the program itself appears to be a ladder logic diagram, with switch and relay symbols, there are no actual switch contacts or relay coils operating inside the PLC to create the logical relationships between input and output.

译文：尽管这个带有开关和继电器符号的程序看上去像梯形逻辑图，但 PLC 中并不是使用真实的开关或继电器线圈来建立输入输出间的逻辑关系的。

注释：although 引导让步状语从句，主句中"operating…"为现在分词短语作定语修饰"actual switch contacts or relay coils"。

[2] Any and all X1 contacts appearing in the program will assume the actuated (non-normal) state, as though they were relay contacts actuated by the energizing of a relay coil named "X1".

译文：程序中所有 X1 的触点都将呈现激活（非常态）状态，它们就像继电器的触点那样被名为 X1 的继电器线圈激活。

注释：appearing in the program 为现在分词短语作定语修饰主语"Any and all X1 contacts"。Though 引导状语从句，谓语用虚拟语气"were"，表示与事实相反，"actuated by…"部分为过去分词短语作定语修饰"relay contacts"。

[3] Equally important to understand is that the personal computer used to display and edit the PLC's program is not necessary for the PLC's continued operation.

译文：同样重要的是，我们要明白用来显示和编辑 PLC 程序的个人电脑对于 PLC 的持续运行并不是必要的。

注释：该句中含一个名词性 that 从句用作表语，该从句中"used"部分为过去分词短语作定语修饰从句主语"personal computer"。

[4] I include the personal computer display in these illustrations for your sake only, in aiding to understand the relationship between real-life conditions (switch closure and lamp status) and the program's status ("power" through virtual contacts and virtual coils).

译文：为了帮助理解实际环境（开关的闭合和电灯的状态）和程序状态（虚拟触点和线圈的通电情况）间的关系，在图中加入了电脑的显示器部分。

注释："in aiding to"表示"为了"的意思。

[5] Using a PLC, however, we can program as many contacts as we wish for each "X" input without adding additional hardware, since each input and each output is nothing more than a single bit in the PLC's digital memory (either 0 or 1), and can be recalled as many times as necessary.

译文：可是用 PLC 的话，就可在程序中为输入 X 编制任意多个触点而不需要增加额外的

111

硬件，因为在 PLC 的数字存储器（0 或 1）中每个输入和输出只不过占一个位，并且可以被任意调用。

注释："Using"为现在分词置句首作条件状语从句，主句中含有一个"since"引导的原因状语从句"since each…necessary"。

✎ Exercises

Ⅰ. Mark the following statements with T (true) or F (false) according to the text.

1. The actual logic of the control system is established inside the PLC by means of a computer program. ( )

2. The contact, coil, connecting wires, and "power" appearing in the PLC's programming are all actual. ( )

3. The personal computer used to display and edit the PLC's program is necessary for the PLC's continued operation. ( )

4. One of the advantages of implementing logical control in software rather than in hardware is that input signals can be re-used as many times in the program as is necessary. ( )

5. One should always consider the effects of failed (open) wiring on the device or devices being controlled. ( )

Ⅱ. Complete the following sentences

1. Although the program of PLC appears to be a _____, with switch and relay _____, there are no actual switch _____ operating inside the PLC to create the logical relationships between input and output.

2. Since the PLC is _____, we can alter its behavior by changing _____ we give it, without having to _____ the electrical components connected to it.

3. _____ is just as important in PLC-controlled systems as it is in _____-controlled systems.

4. Since each output in the PLC is nothing more than _____ in its _____ as well, we can assign _____ in a PLC program "actuated" by an _____.

5. As computers, PLCs can perform _____ (for the equivalent of time-delay relays), _____, and other advanced functions with far greater _____ and _____ than what is possible using electromechanical logic devices.

Reading ⑬ Ladder Diagram

Introduction

Programmable controllers are generally programmed in ladder diagram (or "relay diagram") which is nothing but a symbolic representation of electric circuits. Symbols were selected that actually looked similar to schematic symbols of electric devices, and this has made

Chapter 3 Industry Computer Control Technology

it much easier for electricians to switch to programming PLC controllers. Electrician who has never seen a PLC can understand a ladder diagram.

Ladder diagram

There are several languages designed for user communication with a PLC, among which ladder diagram is the most popular. Ladder diagram consists of one vertical line found on the left hand side, and lines which branch off to the right. Line on the left is called a "bus bar", and lines that branch off to the right are instruction lines. Conditions which lead to instructions positioned at the right edge of a diagram are stored along instruction lines. Logical combination of these conditions determines when and in what way instruction on the right will execute. Basic elements of a relay diagram can be seen in Fig.3.15.

Most instructions require at least one operand, and often more than one. Operand can be some memory location, one memory location bit, or some numeric value number. In the example above, operand is bit 0 of memory location IR000. In a case when we wish to proclaim a constant as an operand, designation # is used beneath the numeric writing (for a compiler to know it is a constant and not an address.)

Based on the picture in Fig.3.15, one should note that a ladder diagram consists of two basic parts: the left section also called conditional, and the right section which has instructions. When a condition is fulfilled, instruction is executed, and that's all!

Picture in Fig.3.16 represents a example of a ladder diagram where relay is activated in PLC controller when signal appears at input line 00. Vertical line pairs are called conditions. Each condition in a ladder diagram has a value ON or OFF, depending on a bit status assigned to it. In this case, this bit is also physically present as an input line (screw terminal) to a PLC controller. If a key is attached to a corresponding screw terminal, you can change bit status from a logic one status to a logic zero status, and vice versa. Status of logic one is usually designated as "ON", and status of logic zero as "OFF".

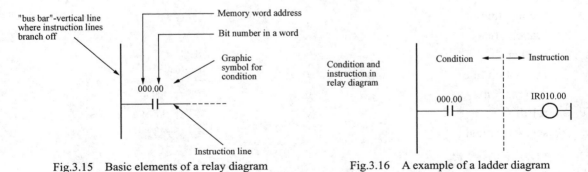

Fig.3.15 Basic elements of a relay diagram   Fig.3.16 A example of a ladder diagram

The right section of a ladder diagram is an instruction which is executed if left condition is fulfilled. There are several types of instructions that could easily be divided into simple and complex. Example of a simple instruction is activation of some bit in memory location. In the example above, this bit has a physical connotation because it is connected with a relay inside a PLC controller. When a CPU activates one of the leading four bits in a word IR010, relay contacts move and connect lines attached to it. In this case, these are the lines connected to a screw terminal marked as 00 and to one of COM screw terminals.

Brief example

Example below represents a basic program. Example consists of one input device and one output device linked to the PLC controller output. Key is an input device, and a bell is an output supplied through a relay 00 contact at the PLC controller output. Input 000.00 represents a condition in executing an instruction over 010.00 bit. Pushing the key sets off a 000.00 bit and satisfies a condition for activation of a 010.00 bit which in turn activates the bell. For correct program function another line of program is needed with END instruction, and this ends the program.

The picture in Fig.3.17 depicts the connection scheme for this example.

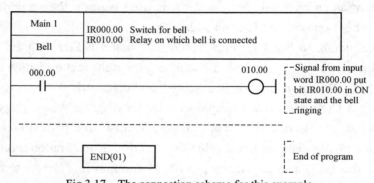

Fig.3.17 The connection scheme for this example

Technical Words and Expressions

vertical line		垂直线
branch off		分叉
bus bar		母线
instruction line		指令线
designation	n.	指示，指定
fulfill	vt.	实现，完成
screw terminal		螺纹接线端子

Comprehension

1. Why Electrician who has never seen a PLC can understand a ladder diagram? _____.

    A. A ladder diagram is a kind of assembler language

B. A ladder diagram looks like actual circuit

C. Symbols used in a ladder diagram look like actual electric devices

D. Symbols used in a ladder diagram look like schematic symbols of electric devices

2. Which of the following is right? _____.

   A. PLCs are programmed in ladder diagram only

   B. A ladder diagram is a kind of electric circuits

   C. A ladder diagram is used to communication with a PLC

   D. Ladder diagram consists of a "bus bar" and a instruction line

3. In Fig.3.16 _____.

   A. if the Status of 000.00 is off, the condition is on

   B. if a key is attached to a corresponding screw terminal, you can change bit status from a logic zero status to a logic one status

   C. if a key is attached to a corresponding screw terminal, you can only change bit status from a logic one status to a logic zero status

   D. status of logic one must be designated as "ON", and status of logic zero as "OFF"

4. An instruction _____.

   A. is determined by the condition

   B. is activation of some bits in I/O

   C. may be connected with an actual relay

   D. locates on the left of its condition

5. In Fig.3.17, _____.

   A. the Ladder diagram consists of a "bus bar" and a instruction line

   B. if the switch is off, then IR 010.00 bit will be activated

   C. input device determines the state of IR000.00

   D. END instruction is not necessary

Unit 14 Introduction to Microcontrollers

A microcontroller can be compared to a small stand alone computer, and it is a very powerful device, which is capable of executing a series of preprogrammed tasks and interacting with other hardware devices. Being packed in a tiny integrated circuit (IC) whose size and weight is usually negligible, it is becoming the perfect controller for robots or any machines requiring some kinds of intelligent automation. A single microcontroller can be sufficient to control a small mobile robot, an automatic washer machine or a security system. Any microcontroller contains a memory to store the program to be executed, and a number of input/output lines that can be used to interact with other devices.

Nowadays, microcontrollers are so cheap and easily available that it is common to use them instead of simple logic circuits like counters for the sole purpose of gaining some design flexibility and saving some space[1]. Some machines and robots will even rely on a multitude of microcontrollers, each one dedicated to a certain task. Most recent microcontrollers are "In System Programmable", meaning that you can modify the program being executed, without removing the microcontroller from its place.

The 8051 microcontroller architecture

The 8051 is the name of a big family of microcontrollers. The device which we are going to use is the "AT89S52" (Fig.3.18) which is a typical 8051 microcontroller manufactured by AtmelTM[2].

Fig.3.18 AT89S52

A simpler architecture of 89S52 can be represented as the Fig.3.19. Fig.3.19 shows the main features and components that the designer can interact with. You can notice that the 89S52 has 4 different ports, each one having 8 input/output lines providing a total of 32 I/O lines. Those ports can be used to output DATA and orders do other devices, or to read the state of a sensor, or a switch. Most of the ports of the 89S52 have "dual function", meaning that they can be used for two different functions: the fist one is to perform input/output operations and the second one is used to implement special features of the microcontroller like counting external pulses, interrupting the execution of the program according to external events, performing serial data transfer or connecting the chip to a computer to update the software. Each port has 8 pins, and will be treated from the software point of view as an 8-bit variable called "register", each bit being connected to a different input/output pin[3].

You can also notice two different memory types: RAM and EEPROM. Shortly, RAM is used to store variables during program execution, while the EEPROM memory is used to store the program itself, that's why it is often referred to as the "program memory".

The special features of the AT89S52 microcon troller are grouped in the blue box at the bottom of Fig.3.19.

It is clear that the CPU (Central Processing Unit) is the heart of microcontrollers. It is the CPU that will Read the program from the FLASH memory and execute it by interacting with the different peripherals discussed above.

Fig.3.20 shows the pin configuration of the 89S52, where the function of each pin is written next to it, and, if it exists, the dual function is written between brackets. The pins are written in the same order as in the block diagram of Fig.3.20, except for the VCC and GND pins which I usually note at the top and the bottom of any device. Note that the pin that has dual functions, can still be used normally as an input/output pin.

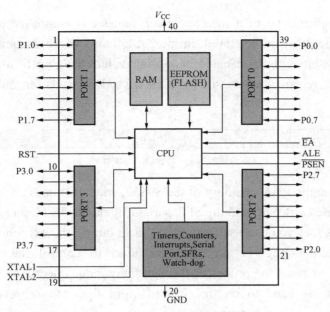

Fig.3.19　Architecture of AT89S52

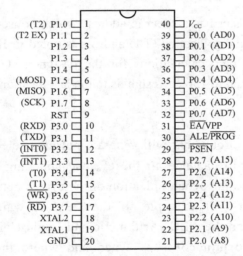

Fig.3.20　The pin configuration of the AT89S52

Memory organization

A RAM stands for Random Access Memory, it has basically the same purpose of the RAM in a desktop computer, which is to store some data required during the execution time of different programs. While an EEPROM, also called FLASH memory is a more elaborated ROM (Read Only Memory) which is the memory where the program being executed is stored[4]. The EEPROM term stands for Electronically Erasable and Programmable Read Only Memory.

In microcontrollers, like in any digital system, memory is organized in Registers, Which is the basic unit of construction of a memory. Each register is composed of a number of bits (usually 8) where the data can be stored. In the 8051 family of microcontrollers for example, most registers are 8-bit register, capable of storing values ranging from 0 to 255. Fig.3.21 shows a typical 8-bit registers, where the notation D0 to D7 stands for the 8 DATA bits of the register.

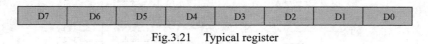

Fig.3.21   Typical register

As you shall see, the RAM memory of the 89S52, which contains 256 registers, is divided into to main parts, the GPR part, and the SFR part. GPR stands for "General Purpose Register" and are the registers that you can use to store any data during the execution of your program. SFRs (Special function Register) are registers used to control the functioning of the microcontroller and to assist the processor through the various operations being executed. For example, SFRs can be used to control Input/Output lines, to retrieve data transmitted through the serial port of a desktop computer, or to configure one of the on-chip counters and timers.

In a memory each register has a specific address which is used by the processor to read and write from specific memory location. The address is noted in Hexadecimal format as this notation simplifies digital logic calculations for the designers, 00 corresponds to the first location and FF which is equal to 256 corresponds to the last location.

Clock concept

The clock is a simple circuit that will generate pulses of electricity at a very specific frequency. Those pulses will cadence all the events happening inside a microcontroller, those pulses will also assure the synchronization of the events between various components inside the microcontroller. For example, if the CPU is waiting for some result of mathematical operation from the ALU (Arithmetic and Logic Unit), it will be known —— according to very specific protocol —— when and where the resulting data will be delivered to the CPU. The synchronization of those two devices is maintained because they share the same clock.

The clock has another very important role which is to enable the microcontroller to count timing. Without a precise clock, it would be impossible to build a "Real Time System", or any other device that relies on time measurements. It can be deduced that the precision of the timing of a microcontroller depends on the frequency of its clock.

In the 89S52 microcontroller, the clock can be fixed to different value by connecting a crystal to the pins 18 and 19. The maximum operating frequency of the AT89S52 is 33 MHz.

Chapter 3 Industry Computer Control Technology

Technical Words and Expressions

microcontroller			微控制器
preprogrammed		adj.	预编程序的
negligible	['neglidʒəbl]	adj.	可以忽略的，不予重视的
intelligent	[in'telidʒənt]	adj..	[计] 智能的
security system			安全系统
multitude	['mʌltitju:d]	n.	大量；众多
dedicate to			献身，用于
architecture	['ɑ:kitektʃə]	n.	体系机构，结构；建筑
sensor	['sensə]		传感器
interrupt	[,intə'rʌpt]	n.	中断，中断信号
RAM (Random-access memory)			随机存储器
EEPROM			电可擦除只读存储器
FLASH memory			闪存
peripheral	[pə'rifərəl]	n.& adj.	外围设备；外围的
bracket	['brækit]	n.	括弧
elaborate	[i'læbərət]	adj.	精致的，精巧的
notation	[nəu'teiʃən]	n.	符号
GPR (General Purpose Register)			通用寄存器
SFR			特殊功能寄存器
serial port			[计] 串行端口
on-chip			芯片内
memory location			存储单元
cadence	['keidəns]	v.	节奏，韵律
synchronization	[,siŋkrənai'zeiʃən]	n.	同步
ALU (Arithmetic and Logic Unit)			算术逻辑部件，运算器
real time system			实时系统
fix	[fiks]	vi.	选定，确定
crystal	['kristl]	n.	石英晶振

Notes

[1] Nowadays, microcontrollers are so cheap and easily available that it is common to use them instead of simple logic circuits like counters for the sole purpose of gaining some design flexibility and saving some space.

译文：现在微控制器/单片机非常便宜且容易买到，因此为了增强设计的灵活性及节省空间，常用它们来替代一些像计数器这样简单的逻辑电路。

注释：此句为 so...that... 的结构。

[2] The device which we are going to use is the "AT89S52" which is a typical 8051 microcontroller manufactured by Atmel TM.

译文：我们将使用的芯片是 AT89S52，它是 ATRNEL 公司制造的典型的 8051 系列微控制器/单片机。

注释：该句结构为：The device（主语）＋is（谓语）＋the "AT89S52"（宾语），其中第一个 which 定语从句修饰 "device"，第二个 which 定语从句修饰 "AT89S52"，该从句中又有一个过去分词短语（manufactured by Atmel TM）作定语修饰 "8051 microcontroller"。

[3] Each port has 8 pins, and will be treated from the software point of view as an 8-bit variable called "register", each bit being connected to a different Input/Output pin.

译文：每个端口有 8 个引脚，在软件中它们被看作一个 8 位的变量，称为寄存器，每一位都被连到不同的输入/输出引脚上。

注释：and 引导两个主语相同的并列句，第二句的主语省略，其宾语 "8-bit variable" 带一个过去分词短语作定语 called "register"。

[4] While an EEPROM, also called FLASH memory is a more elaborated ROM (Read Only Memory) which is the memory where the program being executed is stored.

译文：而 EEPROM 又被称为闪存是一个更为复杂的 ROM（只读存储器），它是个用来存储被执行程序的存储器。

注释：which 引导定语从句修饰 "ROM"，该从句中又包含了一个 where 引导的定语从句修饰 "memory"，而在 where 从句中 "being executed" 为现在分词作定语修饰 "the program"。

Exercises

Ⅰ. Mark the following statements with T (true) or F (false) according to the text.

1. Nowadays, microcontrollers are cheap and easily available. （ ）
2. The 8051 is the name of a big family of microcontrollers. （ ）
3. AT89S52 has 3 different ports, each one having 8 Input/output lines providing a total of 24 I/O lines. （ ）
4. There are two different memory types in AT89S52: RAM and EEPROM. （ ）
5. The ROM memory of the AT89S52 contains 256 registers. （ ）

Ⅱ. Complete the following sentences.

1. Most recent microcontrollers are '_____', meaning that you can _____ the program _____, without removing the microcontroller from its place.

2. Most of the ports of the AT89S52 have '_____' meaning that they can be used for two different _____: the fist one is to perform input/output operations and the second one is used to _____ of the microcontroller like counting external pulses, _____ the execution of the program according to external events, performing _____.

3. It is the CPU that will Read the program from the _____ and execute it by interacting

Chapter 3 Industry Computer Control Technology

with _____.

4. A RAM stands for _____, which is to store some _____ required during the execution time of different programs. While an EEPROM, also called _____ is a more elaborated ROM (_____) which is the memory where the program being executed is stored.

5. The _____ is a simple circuit that will generate _____ of electricity at a very specific _____. In the 89S52 microcontroller, the _____ can be _____ to different value by connecting _____ to the pins 18 and 19.

Reading 14 The Beginning

As always, beginning is the most difficult. You have bought microcontroller, you have learned everything about its systems and registers, you have great idea how to apply all that in practice. The only thing left over to you is to start...

How to start working?

Microcontroller is a good-natured "giant from the bottle" and there is no need for extra knowledge in order to use it.

In order to create your first device under microcontroller's control, you need: the simplest PC, program for compiling into machine code and simple device for "transferring" that code from PC to chip itself.

The process itself is quite logical but dilemmas are anyway common, not because it is complicated but for the fact that there are numerous variations. Let's start...

Writing program in assembler

In order to write a program for the microcontroller, a specialized program in Windows environment may be used. It may, but it does not have to... On using such a software, there are numerous tools which facilitate operating (first of all simulator tool). This is a distinct advantage, but there are other ways too. Basically, text is the only important thing. Because of that any program for text processing can be used for writing program. The point is to write all instructions in order they should be executed by the microcontroller. The rules of assembly language are observed and instructions are written as they are defined. Program idea is followed. That's all!

```
        RESET       VECTOR
                    CSEG      AT      0
                    JMP       XRESET              ; Reset vector
                    CSEG
                    ORG       100H
        XRESET:     ORL       WMCON, #PERIOD      ; Defining of Watch-dog period
                    ORL       WMCON, #WDTEN       ; Watch-dog timer is enabled
```

If a document is written for being used further by programmer then it has to have an extension, .asm in its name, for example: Program.asm.

If a program is written using a specialized program, this extension will be automatically added. If any other program for text processing (Notepad) is used then the document should be saved and additionally renamed. For example: Program.txt→Program.asm. This procedure is not necessary. The document may be saved in original format while its text may be copied to programmer for further use.

As seen, text of the program is the only thing that matters.

Compiling into machine code

Microcontroller "does not understand" assembly language. That is why this program should be compiled. If a specialized program is used —— nothing simpler —— a machine code compiler is a part of the software! Problem is solved by a click on the appropriate icon. The result is a new document which has extension .hex in its name. That is the same program you have already written, but compiled into machine language which microcontroller perfectly understands. The common name of this document is "hex code" and represents apparently meaningless series of numbers in hexadecimal numerical system, for example:

03000000020100FA1001000075813F;
7590FFB29012010D80F97A1479D40;
90110003278589EAF3698E8EB25B;
A585FEA2569AD96E6D8FED9FAD.

In case some other software for writing program in assembler is used, a software especially installed for compiling into machine code must be used. This compiler is activated, document with extension .asm is open and the appropriate command is executed. The result is the same —— a new document with extension .hex. The only problem is that it is stored in your PC.

Copy program to a microcontroller

A cable for serial communication and a special device called programmer are necessary to transfer "hex code" to the microcontroller. There are also several options on how to do it.

A great deal of programs and electronic circuits having this purpose can be found on Internet. Do as follows: open hex code document, adjust a few parameters and click on the icon for compiling. After a while, a series of zeros and units will be programmed into the microcontroller through the serial connection cable. The only thing left over is to transfer programmed chip to the final device. In case it is necessary to change something in the program, the previous procedure may be repeated.

The following pictures show steps of the beginning (Fig.3.22).

Chapter 3 Industry Computer Control Technology

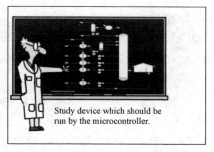

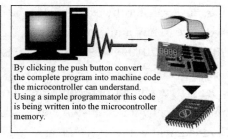
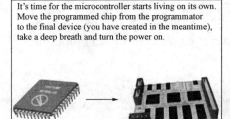

Fig.3.22 Programming for the beginners

Technical Words and Expressions

compile	vt.	编译，编辑，汇编
machine code		［计］机器代码
dilemma	n.	困境
assembler	n.	汇编程序
facilitate	vt.	使容易，使便利，推动
simulator tool		模拟器，仿真工具

assembly language		[计]汇编语言
procedure	n.	程序
hex code		十六进制代码

Comprehension

1. To write a program for the microcontroller, _____.
   A. you must have a specialized software which has numerous tools such as simulator tool
   B. you must have a program for text processing
   C. you must save the program as extension document
   D. all above
2. To compile the assembly language into machine code, _____.
   A. we can't do it without a specialized software which is used to program
   B. we have to writer a totally different document which has extension .hex in its name
   C. we must transfer the extension of the document from .asm to .hex
   D. we have to buy a microcontroller
3. 7590FFB29012010D80F97A1479D40 is _____.
   A. assembly language
   B. a language which Microcontroller "does not understand"
   C. in document with extension .hex
   D. in document with extension .asm
4. To transfer "hex code" to the microcontroller, _____.
   A. we must have a cable for serial communication
   B. we should open hex code document and click on the icon for compiling
   C. we must have a programmer
   D. all above
5. If you want to change the program in a microcontroller, _____.
   A. you must change the hardware of the microcontroller
   B. you have to: write program in assembler, compile the program into machine code and copy machine code to a microcontroller
   C. you have to change the ASM file in the microcontroller
   D. all above

Unit 15 8051 Instruction Set

Introduction

Writing program for the microcontroller mainly consists of giving instructions (commands)

Chapter 3 Industry Computer Control Technology

in that order in which they should be executed later in order to carry out specific task[1]. As electronics can not "understand" what for example instruction "if the push button is pressed-turn the light on" means, then a certain number of simpler and precisely defined orders that decoder can recognize must be used. All commands are known as INSTRUCTION SET. All microcontrollers compatible with the 8051 have in total of 255 instructions, i.e. 255 different words available for program writing.

At first sight, it is imposing number of odd signs that must be known by heart. However, it is not so complicated as it looks like. Many instructions are considered to be "different", even though they perform the same operation, so there are only 111 truly different commands. For example: ADD A, R0, ADD A, R1... ADD A, R7 are instructions that perform the same operation (addition of the accumulator and register) but since there are 8 such registers, each instruction is counted separately! Taking into account that all instructions perform only 53 operations (addition, subtraction, copy etc.) and most of them are rarely used in practice, there are actually 20~30 shortened forms needed to be known, which is acceptable.

Types of instructions

Depending on operation they perform, all instructions are divided in several groups:
- Arithmetic Instructions;
- Branch Instructions;
- Data Transfer Instructions;
- Logical Instructions;
- Logical Instructions with bits.

The first part of each instruction, called MNEMONIC refers to the operation an instruction performs (copying, addition, logical operation etc.). Mnemonics commonly are shortened form of name of operation being executed. For example:

INC R1—means: Increment R1 (increment register R1).

Another part of instruction, called OPERAND is separated from mnemonic at least by one empty space and defines data being processed by instructions[2]. Some instructions have no operand, some have one, two or three. If there is more than one operand in instruction, they are separated by comma. For example:
- RET—(return from sub-routine)—no operand;
- ADD A, R3—(add R3 and accumulator)—two operand;
- CJNE A, #20, LOOP—(compare accumulator with 20. If they are not equal, jump to address specified as LOOP)—three operand.

The operands listed below are written in shortened forms having the following meaning:
- A—Accumulator;
- Rn—Rn is one of R registers (R0~R7) in the currently active bank in RAM;
- Rx—Rx is any register in RAM with 8-bit address. It can be a general-purpose register

or SFR Register (I/O port, control register etc.);
- @Ri—Ri is R0 or R1 register in the currently active bank. It contains register;
- address—the instruction is referring to;
- #X—X is any 8-bit number (0~255);
- #X16—X is any 16-bit number (0~65535);
- adr16—16-bit address is specified;
- adr11—11-bit address is specified;
- rel—The address of a close memory location is specified (−128 to +127 relative to the current one);
- bit—Bit address is specified;
- C—Carry bit in the status register (register PSW).

Arithmetic instructions

These instructions perform several basic operations (addition, subtraction, division, multiplication etc.). After execution, the result is stored in the first operand. For example:

ADD A, Rn—Instruction adds the number in the accumulator and the number in register Rn (R0~R7). After addition, the result is stored in the accumulator.

SUBB A,Rx

Description: Instruction performs subtract operation: A-Rx including the Carry bit as well which acts as borrow. If the higher bit is subtracted from the lower bit then the Carry bit is set. As it is direct addressing, Rx can be some of SFRs or general-purpose register with address 0~7FH. (0~127dec.). The result is stored in the accumulator.

DEC Rx

Description: Instruction decrements value of the register Rx by 1. As it is direct addressing, Rx must be within the first 255 locations of RAM. If there is a 0 in the register, the result will be FFH.

Branch instructions

There are two kinds of these instructions:

Unconditional jump instructions: after their execution a jump to a new location from where the program continues execution is executed[3]. For example:
- ACALL adr11—Call subroutine located at address within 2K byte Program Memory space.
- RET—Return from subroutine.

Conditional jump instructions: if some condition is met—a jump is executed. Otherwise, the program normally proceeds with the next instruction. For example:

JBC bit, rel

Description: This instruction first check if the bit is set. If it is set, a jump to the specified address is executed and afterwards the bit is cleared. Otherwise, the program proceeds with the first next instruction. This is a short jump instruction, which means that the address of a new location must be relatively near the current position in the program (−129 to + 127 locations relative to the first following instruction).

Data transfer instructions

These instructions move the content of one register to another one. The register which content is moved remains unchanged. If they have the suffix "X" (MOVX), the data is exchanged with external memory. For example:

MOV A,@Ri

Description: Instruction moves the Rx register to the accumulator. Rx register address is stored in the Ri register (R0 or R1). After instruction execution, the result is stored in the accumulator. The Rx Register is not affected.

XCH A, Rx

Description: Instruction sets the contents of the accumulator into the register Rx. At the same time, the content of the Rx register is set into the accumulator. As it is direct addressing, the register Rx can be some of SFRs or general-purpose register with address 0~7FH (0~127 dec.).

Logical instruction

These instructions perform logical operations between corresponding bits of two registers. After execution, the result is stored in the first operand.

ADD A, Rx

Description: Instruction adds the accumulator and Rx register. As it is direct addressing, Rx can be some of SFRs or general-purpose register with address 0~7 FH. The result is stored in the accumulator.

ORL A, #X

Description: Instruction performs logical-OR operation between the accumulator and number X. The result of this logical operation is stored in the accumulator.

CLR A/C

Description: Clear accumulator/Carry bit.

Logical operations on bits

Similar to logical instructions, these instructions perform logical operations. The difference is that these operations are performed on single bits.

ANL C, bit

Description: Instruction performs logical-AND operation between the addressed bit and Carry bit.

Technical Words and Expressions

execute	[ˈeksikjuːt]	vt.	执行，实行
instruction set			指令组（表），指令系统
compatible	[kəmˈpætəbl]	adj.	谐调的，一致的，兼容的
addition	[əˈdiʃən]	n.	加法
subtraction	[səbˈtrækʃən]	n.	[数学]减法
arithmetic instruction			算术运算指令
branch instruction			分支指令
data transfer instruction			数据传送指令
logical instruction			逻辑操作指令
logical instructions with bit			位操作逻辑指令
mnemonic	[ni(ː)ˈmɔnik(əl)]	n.	操作码
operand	[ˈɔpəˌrænd]	n.	[计]操作数
comma	[ˈkɔmə]	n.	逗点，逗号
subroutine	[ˌsʌbruːˈtiːn]	n.	[计]子程序
accumulator	[əˈkjuːmjuleitə]	n.	累加器，加法器
jump	[dʒʌmp]	n.	[计]跳转
general-purpose register			通用寄存器
status register			状态寄存器
division	[diˈviʒən]	n.	除法
external memory			外存储器
direct addressing			直接寻址

Notes

[1] Writing program for the microcontroller mainly consists of giving instructions (commands) in that order in which they should be executed later in order to carry out specific task.

译文：对微处理器的编程主要是按顺序给出指令，然后执行指令，从而完成某一特定任务。

注释：该句的主语为"Writing program for the microcontroller"，动名词"writing"作主语。in which 引导一个定语从句，修饰"order"。

[2] Another part of instruction, called OPERAND is separated from mnemonic at least by one empty space and defines data being processed by instructions.

译文：指令的另一部分叫作操作数，它是通过至少一个空格与操作码进行隔离并对指令中处理的数据进行定义。

注释："and"引导两个主语相同的并列句，后一句省略主语，"being processed by instructions"为现在分词短语作定语修饰"data"。

[3] Unconditional jump instructions: after their execution a jump to a new location from where the program continues execution is executed.

译文：无条件跳转指令：当程序执行完该指令，程序将跳转到一个新的地址继续执行程序。

注释：该句的主干为：a jump（主语）＋is executed（谓语）。其中，from where 引导一个定语从句。

Exercises

Ⅰ. **Mark the following statements with T (true) or F (false) according to the text.**

1. All microcontrollers compatible with the 8051 have in total of 111 instructions.　　（　　）
2. Many instructions are considered to be "same", if they perform the same operation.　　（　　）
3. Some instructions have no operand, some have one, two or three.　　（　　）
4. Arithmetic instructions perform several basic operations. After execution, the result is stored in the last operand.　　（　　）
5. Logical Instructions are performed on single bits.　　（　　）

Ⅱ. **Complete the following sentences**

1. Depending on operation they perform, all instructions are divided in several groups: _____, _____, _____, _____ Logical Instructions with bits

2. The first part of each instruction, called _____ refers to _____. Another part of instruction, called _____ is separated from mnemonic at least by one _____ and defines data being processed by instructions.

3. ADD A, Rn - Instruction adds the number in _____ and the number in _____ Rn (R0-R7). After addition, the result is stored in _____.

4. There are two kinds of _____: Unconditional jump instructions and _____.

5. XCH A, Rx - Instruction sets the contents of the _____ into the register Rx. At the same time, the content of the Rx register is set into the accumulator. As it is _____, the register Rx can be some of SFRs or _____ with address 0～7Fh (0～127 dec.).

Reading 15　Basic I/O Operations of 8051

I/O port detailed structure

It is important to have some basic notions about the structure of an I/O port in the 8051 architecture. You will notice along this tutorial how this will affect our choices when it comes to connect I/O devices to the ports. Actually, the I/O ports configuration and mechanism of the 8051 can be confusing, due to the fact that a pin acts as an output pin as well as an input pin in the same time.

Fig.3.23 shows the internal diagram of a single I/O pin of port 1. The first thing you have to notice is that there are tow different direction for the data flow from the microcontroller's processor and the external pin: The Latch value and the Pin value. The latch value is the value that the microcontroller tries to output on the pin, while the pin value, is the actual logic state of the pin, regardless of the latch value that was set by the processor in the first place. The microcontroller reads the state of a pin through the Pin value line, and writes through the latch value line. If you imagine the behavior of the simple circuit in Fig.3.23, you'll notice that the I/O pin should follow the voltage change of the Latch value, providing 5V through the pull-up resistor, or 0V by connecting the pin directly to the GND through the transistor.

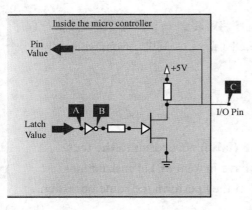

Fig.3.23 Basic I/O pin internal diagram

When the pin is pulled high by the pull-up resistor, the pin can output 5V but can also be used as an input pin, because there is no any risk of short-circuit due to the presence of a resistor. This can be easily verified by connecting the pin to 0V or to 5V, the tow possible outcomes are both unharmful for the microcontroller, and the PIN value line will easily follow the value imposed by the external connection.

Now imagine the opposite configuration, where the latch value would be low, causing the pin to provide 0V, being directly connected to GND through the resistor. If in this situation, an external device tries to raise the pin's voltage to 5V, a short circuit will occur and some damage may be made to the microcontroller's port or to the external device connected to that pin.

To summarize, in the 8051 architecture, to use a PIN as an input pin, you have to output "1", and the pin value will follow the value imposed by the device connected to it (switch, sensor, etc.). If you plan to use the pin as an output pin, then just output the required value without taking any of this in consideration.

Even if some ports like P3 and P0 can have a slightly different internal composition than P1, due to the dual functions they assure, understanding the structure and functioning of port 1 as described above is fairly enough to use all the ports for basic I/O operations.

A simple output project: Blinking a led

A first simple project to experiment with the output operations is to blink a LED. Assuming you have successfully written and compiled the code as explained in the previous part of the tutorial, now we are going to transfer the HEX file corresponding to that code on the 89S52 microcontroller. Let us recall that the HEX file is a machine language file, generated by the compiler, originally from a C code.

The code for blinking a LED is as follow:

```
#include <REGX52.h>
#include <math.h>
delay(unsigned int y){
unsigned int i;
   for(i=0;i<y;i++){;}
}
main(){
   while(1){
   delay(30000);
   P1_0 = 0;
   delay(30000);
   P1_0 = 1;
   }
}
```

Before transferring the HEX file to the target microcontroller, the hardware have to be constructed. First you have to provide a clean (noiseless) 5V power supply, by connecting the Vcc pin (40) to 5V and the GND pin (20) to 0V. Then you have provide a mean of regulating or generating the clock of the processor. The easiest and most efficient way to do this is to add a crystal resonator and 2 decoupling capacitors of approximately 30pF (see the crystal X1 and the capacitors C1 and C2 on Fig.3.23). Then, you have connect pin 31 (EA) to 5V. The EA pin is an active low pin that indicate the presence of an external memory. Activating this pin by providing 0V on it will tell the internal processor to use external memories and ignore the internal built-in memory of the chip. By providing 5V on the EA pin, its functionality is deactivated and the processor uses the internal memories (RAM and FLASH). At last, you have to connect a standard reset circuitry on pin 9 composed of the 10 kohm resistor R2 and the 10μF capacitor C3, as you can see in the schematic. You can also add a switch to short-circuit pin 9 (RST) and 5V giving you the ability to reset the microcontroller by pressing on the switch (the processor resets in a high level is provided on the RST pin for more than 2 machine cycles).

Those were the minimum connections to be made for the microcontroller to be functional and able to operate correctly. According to the fact that we are going to use an ISP programmer, A connector is added by default to allow easy in system programming.

For our simple output project, a LED is connected to P1.0 through a 220 ohm resistor R1, as you can see in Fig.3.24 below. Note that there are other ways to connect the LED, but now that you understand the internal structure of the port, you can easily deduce that this is the only way to connect the LED so that the current is fully controlled by the external resistor R1. Any other connection scheme would involve the internal resistor of the port, which is "uncontrollable".

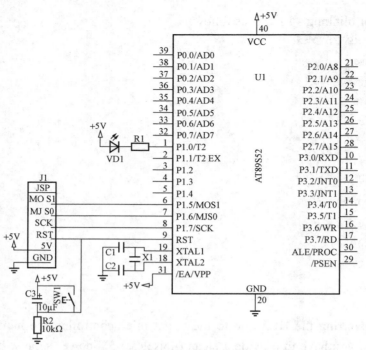

Fig.3.24 LED blinking project hardware

In order to get rid of any confusion, a picture of the implementation of this simple project on a bread board is provided in Fig.3.25 to help you visualize the hardware part of the project:

Note that the reset switch and R/C filter are not present on this breadboard, the reset functionality of the ISP cable was used instead.

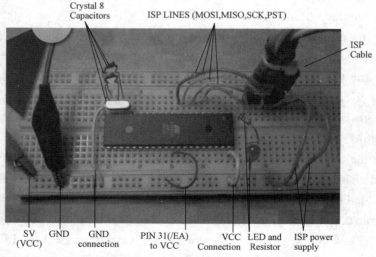

Fig.3.25 LED blinking project hardware implementation

At this stage, you can finally connect your ISP programmer, launch the ISP prog software, browse the HEX file for programming the FLASH, and transfer it to the microcontroller, as described in the ISP page. You can eventually use any other available programming hardware and/or software.

If all your connections are correct, you should see the LED blinking as soon as the programming (transfer) is finished. You can experiment with different delay in the code to change the blinking frequency. Don't forget that for any change to take place, you have to rebuild your source code, generating a new hex file (replacing the old one) and retransfer the freshly generated HEX file to the microcontroller.

Technical Words and Expressions

notion	n.	概念，观念，想法，意见
I/O port		输入/输出端口
tutorial	n.	指南
latch value		锁存值
pull-up resistor		上拉电阻
crystal resonator		晶体谐振器
decoupling capacitor		解耦电容
breadboard	n.	电路试验板，面包板
source code		原始（代）码，源程序

Comprehension

1. In Fig.3.23, if the latch value is high lever, I/O pin will be _____.
   A. 5V B. 0V C. 3V D. 7V
2. For basic I/O operations, the structure and functioning of port 1 is _____ port2.
   A. the same as   B. more complicated than
   C. more simple than   D. easier to understand than
3. In Fig.3.24, the circuit between pin 18 and pin19 _____.
   A. is used to provide reset circuitry to the processor
   B. is used to provide power to the processor
   C. is used to regulate power supply
   D. is used to provide a clock to the processor
4. If the EA pin is in low lever, _____.
   A. the processor will use external memories only
   B. the processor will use internal memories
   C. the processor will use RAM and FLASH
   D. the processor will use both external and internal memories
5. If you want to change the blinking frequency of LED, _____.
   A. you must change the hardware of the project

B. you only have to change the C code of this project
C. you have to change the HEX file in the microcontroller
D. all above

Unit 16 Fieldbus

Fieldbus (or field bus) is the name of a family of industrial computer network protocols used for real-time distributed control, now standardized as IEC 61158.

A complex automated industrial system —— say a manufacturing assembly line ——usually needs an organized hierarchy of controller systems to function. In this hierarchy there is usually a Human Machine Interface (HMI) at the top, where an operator can monitor or operate the system. This is typically linked to a middle layer of programmable logic controllers (PLC) via a non time critical communications system (e.g. Ethernet). At the bottom of the control chain is the fieldbus which links the PLCs to the components which actually do the work such as sensors, actuators, electric motors, console lights, switches, valves and contactors.

Description

Fieldbus is an industrial network system for real-time distributed control. It is a way to connect instruments in a manufacturing plant. Fieldbus works on a network structure which typically allows daisy-chain, star, ring, branch, and tree network topologies. Previously computers were connected using RS-232 (serial connections) by which only two devices could communicate. This would be the equivalent of the currently used 4~20mA communication scheme which requires that each device has its own communication point at the controller level, while the fieldbus is the equivalent of the current LAN-type connections, which require only one communication at the controller level and allow multiple (100's) of analog and digital points to be connected at the same time[1]. This reduces both the length of the cable required and the number of cables required. Furthermore, since devices that communicate through fieldbus require a Microprocessor, multiple points are typically provided by the same device[2]. Some fieldbus devices now support control schemes such as PID control on the device side instead of forcing the controller to do the processing...

History

Although fieldbus technology has been around since 1988, with the completion of the ISA S50.02 standard, the development of the international standard took many years. In 1999, the IEC SC65C/WG6 standards committee met to resolve difference in the draft IEC fieldbus standard. The result of this meeting was the initial form of the IEC 61158 standard with eight different protocol sets called "Types" as follows:
- Type 1 FOUNDATION Fieldbus H1;
- Type 2 ControlNet;

Chapter 3 Industry Computer Control Technology

- Type 3 PROFIBUS;
- Type 4 P-Net;
- Type 5 FOUNDATION Fieldbus HSE (High Speed Ethernet);
- Type 6 SwiftNet (a protocol developed for Boeing, since withdrawn);
- Type 7 WorldFIP;
- Type 8 Interbus.

This form of "standard" was first developed for the European Common Market, concentrates less on commonality, and achieves its primary purpose —— elimination of restraint of trade between nations. Issues of commonality are now left to the international consortia that support each of the fieldbus standard types. Almost as soon as this "8-headed monster" was approved, the IEC standards development work ceased and the committee was dissolved. A new IEC committee SC65C/MT-9 was formed to resolve the conflicts in form and substance within the more than 4000 pages of IEC 61158. The work on the above protocol types is substantially complete. New protocols, such as for safety fieldbuses or realtime ethernet-based fieldbuses are being accepted into the definition of the international fieldbus standard during a typical 5-year maintenance cycle.

Recent additions or planned additions to IEC 61158 include but are not limited to:
- Type 10 PROFINET IO;
- Type 12 EtherCAT;
- Type 13 Ethernet Powerlink;
- Type 16 SERCOS_interface.

Both FOUNDATION Fieldbus and Profibus technologies are now commonly implemented within the process control field, both for new developments and major refits. In 2006, China saw the largest FF systems installations at NanHai and SECCO, each with around 15000 fieldbus devices connected.

Standards

There are a wide variety of concurring fieldbus standards. Some of the most widely used ones include:
- AS Interface;
- CAN;
- Industrial Ethernet;
- Interbus;
- LonWorks;
- Modbus;
- PROFIBUS;
- BITBUS;
- CompuBus;
- SafetyBUS p.

Cost advantage

A major advantage of fieldbus implementation is the capital expenditure (CAPEX) savings associated with cable elimination; multiple devices share wire-pairs in order to communicate over the bus network and savings are also available through speedier commissioning.

Users have now found that ongoing maintenance and process control system performance are also very significantly enhanced through adopting fieldbus systems, which results in operations expense savings (OPEX)[3].

Disadvantages

There are disadvantages to using fieldbus compared to the 4~20mA analog signal standard (or to 4~20mA with HART):

- Fieldbus systems are more complex, so users need to be more extensively trained or more highly qualified.
- The price of fieldbus components is higher.
- Fieldbus test devices are more complex compared to a (high-spec) multimeter that can be used to read and simulate analog 4~20mA signals.
- Slightly longer reaction times with fieldbus, depending on the system.
- Device manufacturers have to offer different versions of their devices (e.g. sensors, actuators) due to the number of different (incompatible) fieldbus standards. This can add to the cost of the devices and to the difficulty of device selection and availability.
- One or more fieldbus standards may predominate in future and others may become obsolete. This increases the investment risk when implementing fieldbus.

Current developments

In recent years a number of Ethernet-based industrial communication systems have been established, most of them with extensions for real-time communication. These have the potential to replace the traditional field buses in the long term. Currently the issue stopping most Ethernet fieldbus implementations is the availability of device power[4]. Most industrial measurement & control devices need to be powered from the bus and Power-over-Ethernet (PoE) does not deliver enough.

Here is a partial list of the new Ethernet-based industrial communication systems:

- EtherCAT;
- ETHERNET/IP;
- Ethernet Powerlink;
- Industrial Ethernet;
- PROFINET IO;
- SafetyNET p;

- SERCOS Ⅲ;
- TTEthernet;
- VARAN.

Market

In process control systems, the market is dominated by FOUNDATION fieldbus and PROFIBUS PA. Both technologies use the same physical layer (2-wire manchester-encoded current modulation at 31.25kHz) but are not interchangeable. As a general guide, applications which are controlled and monitored by PLCs (programmable logic controllers) tend towards PROFIBUS, and applications which are controlled and monitored by a DCS (digital/distributed control system) tend towards FOUNDATION Fieldbus. PROFIBUS technology is made available through Profibus International with headquarters in Karlsruhe, Germany. FOUNDATION Fieldbus technology is owned and distributed by the Fieldbus Foundation of Austin, Texas.

Technical Words and Expressions

fieldbus		n.	现场总线
protocol	['prəutəkɔl]	n.	协议
real time			[计] 实时
distributed control			分布式控制
assembly line			生产线，装配线
hierarchy	['haiərɑːki]	n.	层次，层级
Human Machine Interface (HMI)			人机界面
actuator	['æktjueitə]	n.	执行装置
console	[kən'səul]	n.	[计] 控制台
contactor	['kɔntæktə]	n.	接触器
daisy-chain		n.	菊花式链接
topology	[tə'pɔlədʒi]	n.	拓扑，布局；拓扑学
microprocessor	[maikrəu'prəusesə(r)]	n.	[计] 微处理器
FOUNDATION Fieldbus			基金会现场总线
consortia		n.	公会，银行团
ethernet		n.	以太网
installation	[ˌinstə'leiʃən]	n.	[计] 安装，装置，就职
capital expenditure			资本支出，基本建设费用
commissioning	[kə'miʃəniŋ]	n.	试运转；试车
availability	[əˌveilə'biliti]	n.	可用性，有效性，实用性
obsolete	['ɔbsəliːt]	adj.	荒废的，陈旧的，过时的
interchangeable	[intə'tʃeindʒəb(ə)l]	adj.	可互换的
DCS (distributed control system)			集散式控制系统

Notes

[1] This would be the equivalent of the currently used 4～20mA communication scheme which requires that each device has its own communication point at the controller level, while the fieldbus is the equivalent of the current LAN-type connections, which require only one communication at the controller level and allow multiple (100's) of analog and digital points to be connected at the same time.

译文：这就和目前使用的4～20mA通信方式一样，它要求每个设备在控制器层上都有自己的通信接点，而现场总线和目前的局域网一样，它在控制器层上只需要一个通信接点并且可以同时连接多个（100）模拟和数字接点。

注释：While 作为并列连词，引导并列句。Which 引导限定性定语从句修饰"4～20mA communication scheme"，"that each device…" that 引导一个宾语从句。

[2] Furthermore, since devices that communicate through fieldbus require a Microprocessor, multiple points are typically provided by the same device.

译文：此外，由于通过现场总线进行通信的设备需要一个微处理器，同一设备通常会提供多个接点。

注释：Since 引导原因状语从句，表示"由于"的意思，从句中 that 则引导了一个同谓语从句。

[3] Users have now found that ongoing maintenance and process control system performance are also very significantly enhanced through adopting fieldbus systems, which results in operations expense savings (OPEX).

译文：现在用户发现采用现场总线系统后，日常维护也变得简单了，而且过程控制系统的性能也大大被增强，从而带来运行成本的下降。

注释：That 后带一个宾语从句，该从句的主语是"ongoing maintenance …performance"。Which 引导了一个非限定性定语从句。

[4] Currently the issue stopping most Ethernet fieldbus implementations is the availability of device power.

译文：目前阻碍大多数以太网现场总线实施的问题在于总线无法给设备提供电源。

注释：该句主语为"the issue stopping most Ethernet fieldbus implementations"，其中 stopping…为现在分词作定语修饰"the issue"。

Exercises

Ⅰ. Mark the following statements with T (true) or F (false) according to the text.

1. Fieldbus (or field bus) is used for real-time distributed control. (　)
2. Fieldbus is a way to connect computer in a manufacturing plant. (　)
3. Fieldbus only allows two devices to communicate. (　)
4. The "standard" which was first developed for the European Common Market, concentrates on commonality. (　)

5. Applications which are controlled and monitored by a DCS (digital/distributed control system) are dominated by FOUNDATION Fieldbus.   (   )

II. Complete the following sentences.

1. In the hierarchy of automated industrial system, at the top is usually _____, where an operator can monitor or operate the system. This is typically linked to a middle layer of _____ via a non time critical communications system (e.g. Ethernet). At the bottom of the control chain is the _____.

2. Since devices that communicate through fieldbus require a Microprocessor, _____ are typically provided by the same device. Some fieldbus devices now support _____ -such as PID control on the device side.

3. Both _____ and _____ technologies are now commonly _____ within the process control field, both for new developments and major refits.

4. A major advantage of fieldbus implementation is _____ savings associated with _____; multiple devices share wire-pairs in order to communicate over the bus network and savings are also available through speedier commissioning.

5. In process control systems, the market is dominated by _____ and. Both technologies use the same _____ (2-wire manchester-encoded current modulation at 31.25 kHz) but are not interchangeable.

Reading 16 Profibus

PROFIBUS (Process Field Bus) is a standard for field bus communication in automation technology and was first promoted (1989) by BMBF (German department of education and research). It should not be confused with the PROFINET standard for industrial Ethernet.

The history of PROFIBUS goes back to a publicly promoted plan for an association started in Germany in 1987 and for which 21 companies and institutes devised a master project plan called "field bus". The goal was to implement and spread the use of a bit-serial field bus based on the basic requirements of the field device interfaces. For this purpose, respective company members agreed to support a common technical concept for production and process automation. First, the complex communication protocol Profibus FMS (Field bus Message Specification), which was tailored for demanding communication tasks, was specified. Subsequently in 1993, the specification for the simpler and thus considerably faster protocol PROFIBUS DP (Decentralized Peripherals) was completed. It replaced FMS.

The information technology played a decisive role in the development of the automation technology and changed the hierarchies and structures of offices. It now arrives to the industrial environment and its several sectors, from process and manufacture industries to buildings and logistic systems. The possibility of communication between devices and the use of standardized, open and transparent mechanisms are essential components of today's automation concept.

The communication expands rapidly in the horizontal direction at the field level, as well as in the vertical direction integrating all of the hierarchy levels of a system. According to the characteristics of the application and the maximum cost to reach, a gradual combination of different communication systems like Ethernet, PROFIBUS and AS-Interface are the ideal open networks conditions for industrial processes.

With respect to actuators/sensors the AS-Interface is the perfect data communication system, as the binary data signals are transmitted through an extremely simple and low-cost data bus, together with the 24V DC power supply required to feed those sensors and actuators. Another important feature is that the data are transmitted in cycles, in a very efficient and fast way.

At field level, the peripherals distributed, like I/O modules, transducers, drives, valves and operation panels' work in automation systems, through an efficient, real-time communication system, the PROFIBUS DP or PA. The process data transmission is made in cycles, while alarms, parameters and diagnostics are transmitted only when necessary, in a non-cyclic way.

As to cells, the programmable controllers like the PLCs and the PCs communicate between themselves and so require that large data package be transferred in several and powerful communication functions. Furthermore, the efficient integration to the existing corporate communication systems, such as Intranet, Internet and Ethernet is absolutely mandatory. This need is met by PROFINet protocol.

The industrial communication revolution in the technology plays a vital role in the optimization of the process systems and has made a valuable contribution toward improving the use of resources. PROFIBUS is a central connecting link on the automation of data flow.

The PROFIBUS (Fig.3.26), in its architecture, is divided in three main variants:

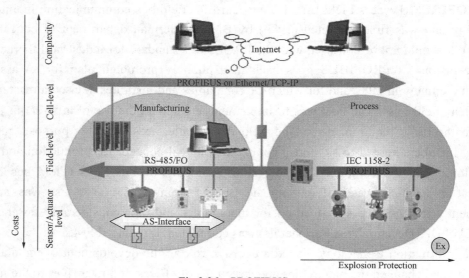

Fig.3.26 PROFIBUS

PROFIBUS-DP: It is the high-speed solution for PROFIBUS. It was developed specifically for communication between automation systems and decentralized equipments. It is applicable on control systems where the access to I/O-distributed devices is emphasized and

Chapter 3 Industry Computer Control Technology

substitutes the conventional 4 to 20mA, HART systems or in 24V transmissions. It uses the RS-485 physical medium or fiber optics. It requires less than two minutes to transmit 1 I/O Kbyte and is largely used in critical time controls.

Currently, 90 percent of the applications involving slaves PROFIBUS utilize the PROFIBUS-DP. This variant is available in three versions: DP-V0 (1993), DP-V1 (1997), DP-V2 (2002). The origin of each version occurred alongside the technological advancement and the long-time growing demand for the applications.

PROFIBUS-FMS: The PROFIBUS-FMS provides the user with a wide selection of functions when compared to other variants. It is the solution in universal communication standard that may be used to solve complex communication tasks between PLCs and DCSs. This variant supports the communication between automation systems besides the exchange of data between intelligent equipments, generally used in control level. Recently, since its primary function is the peer-to-peer communication, it is being replaced by application in Ethernet.

PROFIBUS-PA: The PROFIBUS-PA is the solution PROFIBUS that attend to the requisites of process automation, where automation systems and process control systems connect with field equipments, like pressure and temperature transmitters, converters, positioners, etc. It may also replace the 4 to 20mA standard.

There are potential advantages for using this technology, which carry functional advantages such as the transmission of reliable information, variable status dealing, failure safety systems, and the feature of auto-diagnosis equipment, equipment rangeability, measuring with high resolution, Integration with high speed discreet control, etc. In addition to the economical benefits pertinent to the installations (cost reduction up to 40% comparable to the conventional systems in some cases), maintenance costs reduction (up to 25% against the conventional systems), smaller startup time, it offers a significant increase in functionality and safety.

PROFIBUS-PA permits measurement and control through one line and two single cables. It also powers the field equipment in intrinsically safe areas. PROFIBUS-PA allows maintenance and connecting/disconnecting equipment even during operation without interfering in other stations in areas potentially explosive. PROFIBUS-PA was developed in cooperation with the Control and Process Industry (NAMUR), in compliance with the special requirements on this application area:

- The original application profile for the process automation and interoperability of field equipments from different manufacturers;
- Addition and removal of bus stations even in intrinsically areas without affecting other stations;
- A transparent communication through the couplers of the segment between the PROFIBUS-PA automation bus and the PROFIBUS-DP industrial automation bus.
- Power and data transmission on the same pair of cables based on the IEC 61158-2 technology;
- Use in potentially hazardous areas with "intrinsically safe" or "intrinsically unsafe"

explosive-proof protection shield.

The connection of the transmitters, converters and positioners in a PROFIBUS-DP network is made by a DP/PA coupler. The twisted pair cable is used as power supply and data communication for each equipment, which makes the installation easier and lower the cost of hardware, thus resulting in less initiation time, problem-free maintenance, low engineering software cost and highly reliable operation.

Technical Words and Expressions

promote	vt.	促进，发扬
industrial Ethernet		工业以太网
communication protocol		通信协议
FMS (Field bus Message Specification)		现场总线信息规范
DP (Decentralized Peripheral)		分散的外围设备
transparent	adj.	透明的；透光的
AS-Interface		传感器/执行器接口
transducer	n.	传感器，变换器
substitute	v.	代替，替换，替代
twisted pair		双绞线

Comprehension

1. Which of the following is right? _____.
   A. PROFIBUS standard is a kind of industrial Ethernet
   B. Compared to Profibus FMS, PROFIBUS DP is the simpler and considerably faster protocol
   C. PROFIBUS is the ideal open networks conditions for industrial processes
   D. all above

2. The PROFIBUS DP or PA _____.
   A. is used at cell level
   B. can transfer large data package
   C. can realize real-time data transmission
   D. can realize all data transmission in cycles

3. PROFINet protocol _____.
   A. is a kind of PROFIBUS standard
   B. is a kind of industrial Ethernet
   C. supplies power for sensors and actuators
   D. all above

4. Which of the following is right? _____.

A. PROFIBUS-PA may replace the 4 to 20mA standard
B. PROFIBUS-PA can supply power for the field device
C. maintenance costs of PROFIBUS-PA are cheaper than the conventional systems
D. all above

5. Which of the following is right? _____.
A. PROFIBUS-DP is used for communication among decentralized equipments
B. PROFIBUS-DP has already substituted the conventional 4 to 20mA
C. DP-V1 is a kind of PROFIBUS-DP
D. the applications of PROFIBUS-DP are less than PROFIBUS-FMS

Chapter 4
Power Supply Technology

Chapter 4 Power Supply Technology

Unit 17 Thermal Power Station

A power plant (also known as a power station or power generating station), is an industrial location that is utilized for the generation and distribution of electric power on a mass scale. Many power stations contain one or more generators, a generator is a rotating machine that converts mechanical power into three-phase electric power (these are also known as an alternator)[1]. The relative motion between a magnetic field and an electrical conductor creates an electric current.

Power plants are usually located in suburban areas or far from cities due to their need for large amounts of land and water, as well as requirements for waste disposal.For this reason, a power generating station has to not only concern itself with the efficient generation of power, but also with the transmission of this power. This is why power plants are often closely accompanied by transformer switchyards. These switchyards increase the transmission voltage of the power, which allows it to be more efficiently transmitted over long distances[2].

The energy source used to turn the generator shaft varies and depends on the type of fuel. This fuel choice determines the type of power plant, classifying them accordingly.

Types of power plants

The different types of power plants are classified depending on the type of fuel used. For the purpose of bulk power generation:thermal,nuclear, and hydropower are the most efficient. A power generating station can be broadly classified into the three above-mentioned types.

Thermal power station (Fig. 4.1)

A thermal power station or a coal fired thermal power plant is by far the most conventional method of generating electric power with reasonably high efficiency. It uses coal as the primary fuel to boil the water available to superheated steam for driving the steam turbine.

Fig.4.1 Outside view of thermal power station

The steam turbine is then mechanically coupled to an alternator rotor, the rotation of which results in the generation of electric power[3]. Generally in India, bituminous coal or brown coal are used as fuel of boiler which has volatile content ranging from 8% to 33% and ash content 5% to 16%. To enhance the thermal efficiency of the plant, the coal is used in the boiler in its pulverized form.

In coal fired thermal power plant, steam is obtained at very high pressure inside the steam boiler by burning the pulverized coal. This steam is then super heated in the super heater to extreme high temperature. This super heated steam is then allowed to enter into the turbine, as the turbine blades are rotated by the pressure of the steam.

The turbine is mechanically coupled with alternator in a way that its rotor will rotate with the rotation of turbine blades. After entering into the turbine, the steam pressure suddenly falls, leading to corresponding increase in the steam volume.

After having imparted energy into the turbine rotors, the steam is made to pass out of the turbine blades into the steam condenser of the turbine. In the condenser, cold water at ambient temperature is circulated with the help of a pump which leads to the condensation of the low-pressure wet steam.

Then this condensed water is further supplied to low pressure water heater where the low pressure steam increases the temperature of this feed water, it is again heated at high pressure. This outlines the basic working methodology of a thermal power plant.

Advantages of thermal power plants
- Fuel used i.e coal is quite cheaper.
- Initial cost is less as compared to other generating stations.
- It requires less space as compared to hydro-electric power stations.

Disadvantages of thermal power plants
- It pollutes atmosphere due to production of smoke and fumes.
- Running cost of the power plant is more than hydro electric plant.

Technical Words and Expressions

power plant			发电厂，发电站；动力设备，发电装置
power station			发电站，发电厂
power generating station			发电站，电站
alternator	[ˈɔltərˌneɪtər]	n.	交流发电机
suburban	[səˈbəːbən]	adj.	郊区的，城外的
transformer	[trænsˈfɔːmə(r)]	n.	变压器
switchyard	[ˈswɪtʃjɑːd]	n.	开关站，开关场
generator	[ˈdʒenəreɪtə(r)]	n.	发电机
fuel	[ˈfjuːəl]	n.	燃料
bulk power			大容量电源

thermal	[ˈθɜrməl]	adj.	热的，热量的
nuclear	[ˈnukliər]	adj.	原子能的，核能的，核武器的，核子的
hydropower	[ˈhaɪdroʊˌpaʊə]	n.	水电，水力发电，水能
superheated	[ˌsuːpəˈhiːtɪd]	adj.	过热的
turbine	[ˈtɜːbaɪn]	n.	涡轮机，汽轮机
bituminous	[bɪˈtjuːmɪnəs]	adj.	沥青的，含沥青的
bituminous coal			烟煤
boiler	[ˈbɔɪlə(r)]	n.	锅炉，烧水壶，热水器
volatile	[ˈvɑːlət(ə)l]	adj.	（液体或固体）易挥发的，易气化的
pulverized	[ˈpʌlvəraɪzd]	adj.	粉状的，成粉末的
blade	[bleɪd]	n.	叶片，刀片
condenser	[kənˈdensə(r)]	n.	冷凝器，（尤指汽车发动机内的）电容器
fume	[fjuːm]	n.	烟气，烟雾

Notes

[1] Many power stations contain one or more generators, a generator is a rotating machine that converts mechanical power into three-phase electric power (these are also known as an alternator).

译文：许多发电站都包含一个或多个发电机，这是一种将机械能转换为三相电能的旋转机械（也称为交流发电机）。

注释：句中 that 引导定语从句，对 a rotating machine 也就是 generator 的作用进行说明。

[2] These switchyards increase the transmission voltage of the power, which allows it to be more efficiently transmitted over long distances.

译文：这些开关站提高了电力的传输电压，这使得电力能够远距离更有效地传输。

注释：which 引导非限制性定语从句，对 switchyard 的作用进行了补充说明。

[3] The steam turbine is then mechanically coupled to an alternator rotor, the rotation of which results in the generation of electric power.

译文：然后，蒸汽涡轮机与交流发电机转子机械耦合，交流发电机转子的旋转产生电力。

注释：句中 which 引导定语从句，which 指的是 an alternator rotor。

Exercises

Ⅰ. Mark the following statements with T(true) or F(false) according to the text.

1. Many power stations contain one or more motors, a rotating machine that converts mechanical power into three-phase electric power (these are also known as an alternator).
( )

2. Power plants are usually located in suburban areas or far from cities due to their need

for large amounts of land and water, as well as requirements for waste disposal. ( )

3.These switchyards decrease the transmission voltage of the power, which allows it to be more efficiently transmitted over long distances. ( )

4.A hydropower station is by far, the most conventional method of generating electric power with reasonably high efficiency. ( )

5. In the condenser, cold water at ambient temperature is circulated with the help of a pump which leads to the condensation of the high-pressure wet steam. ( )

Ⅱ. Complete the following sentences.

1. Many power stations contain one or more _____, a rotating machine that converts mechanical power into _____ electric power (these are also known as _____).

2. For the purpose of bulk power generation, thermal、_____ and _____ are the most efficient.

3. The steam turbine is then mechanically coupled to an _____ rotor, the rotation of which results in the _____ of electric power.

4. In coal fired thermal power plant, steam is obtained in very _____ pressure inside the steam _____ by burning the _____ coal.

5. Then this condensed water is further supplied to low pressure water _____ where the low pressure steam increases the _____ of this feed water, it is again heated in_____ pressure.

Reading 17 Components of Thermal Power Plant

In a thermal power plant, various components are used in the cycle. Here we have listed, main components of the thermal power plant(Fig.4.2).

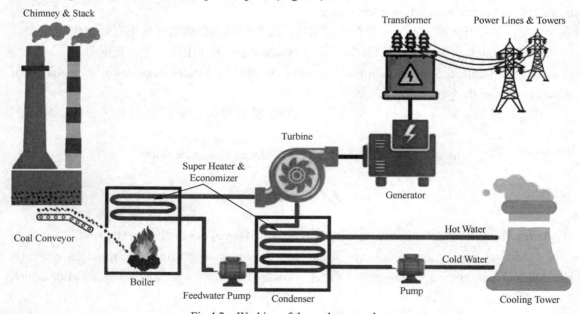

Fig.4.2 Working of thermal power plant

Chapter 4 Power Supply Technology

- Boiler
- Turbine
- Super-heater
- Condenser
- Economizer
- Feedwater pump
- Alternator
- Chimney
- Cooling tower

Boiler

The pulverized coal is fed to the boiler with preheated air. The boiler is used to produce high-pressure steam.

The boiler in the thermal power plant is used to convert the chemical energy of coal into thermal energy or heat energy. During the combustion of coal, a high temperature is produced inside the boiler. This temperature is high enough to convert water into steam.

The size of the boiler depends on the amount of heat required for the thermal power plant. And there are several types of boiler used in a thermal power plant like haycock and wagon top boiler, firetube boiler, cylindrical fire-tube boiler, water-tube boiler, etc.

Turbine

The high pressure and high-temperature steam are fed to the boiler. This superheated steam is a strike on the turbine blade. And the turbine starts rotating. The turbine is a mechanical device that is used to convert the heat energy of steam into rotational energy or kinetic energy.

The turbine is mechanically coupled with an alternator via a shaft. When the steam released from the turbine, the temperature and pressure are reduced. And this steam is passed to the condenser.

Super heater

In a steam turbine, super heated steam is used to rotate the turbine. The wet and saturated steam is supplied to the super heater. And it is a device that converts it into dry and superheated steam.

The super heaters temperature is the highest among all components of the thermal power plant. In the thermal power plant, there are three types of superheaters used: convection, radiant, and separately fired.

The superheater is used to increase the temperature of the steam generated from the boiler. This will increase the thermal energy of the steam.

Condenser

When the steam is released from the turbine, the temperature and pressure are decreased. The exhaust steam of the turbine is reused in the cycle. To increase the turbine efficiency, we need to condense this steam to maintain a proper vacuum.

The condenser decreases the operating pressure. So, the vacuum is increased. And this will increase the volume of steam that results in more amount of work available at the turbine. And due to this, the plant efficiency will increase with the increase in turbine output.

Economizer

The economizer is a heat exchanger device that is used to reduce energy consumption. In the boiler, flue gases are exhausted into the atmosphere. These gases have a high temperature. So, the economizer uses the heat of flue gases to heat the water.

The water released from the condenser is again used in the cycle. With the help of a feedwater pump, this water is transferred to the economizer. An economizer uses the heat of flue gases to increase the temperature of the water.

The economizer uses the waste heat of flue gases. Hence, it is used to increase the efficiency of the entire cycle.

Feedwater pump

A feedwater pump is used to supply water into the boiler. The water may be from the condenser or freshwater. This pump is used to pressure the water. Generally, the feedwater pump is a centrifugal type or positive displacement type of pump.

Alternator

The alternator and turbine are connected on the same shaft. The turbine rotates with the flow of steam and the turbine rotates. The rotor of the alternator rotates and generates electrical energy. Therefore, the alternator is a device that converts kinetic energy or rotational energy into electrical energy.

Chimney

In most of the thermal power plants, coal is used as fuel. During the combustion of coal, the flue gases are generated in the boiler. The chimney provides a path to the flue gas and exhaust to the atmosphere.

The chimney works based on the natural draft and stack effect. The hot air is light in weight and it goes up. The height of the chimney is high. The taller height, the more draft or draught is created.

Cooling tower

As the name suggests, the cooling tower is used to reject waste heat in the atmosphere. Different heat transfer methods are used in the cooling tower. The heat of the water evaporates into the atmosphere. And remains cool water that is further used in the cycle.

The condenser converted steam into water. And the water that comes from the condenser is supplied to the cooling tower. Generally, forced flow cooling towers are used in the thermal power plant. And the air is circulated from bottom to top of the tower.

Technical Words and Expressions

economizer	n.	省煤器
feedwater pump		给水泵
chimney	n.	烟囱
cooling tower		冷却塔
combustion	n.	燃烧，燃烧过程
strike	n.	打击，冲击
convection	n.	对流
radiant	adj.&n.	辐射的；辐射
exhaust steam		废气，废蒸汽
exchanger	n.	交换器，交换机，交换剂
flue gas		烟道气，废气
freshwater	n.	淡水
centrifugal	adj.	离心的
draft	n.	穿堂风
stack	n.&.v.	堆栈，（通常指码放整齐的）一叠，一摞，一大堆，（尤指工厂的）大烟囱；使码放在…，使成叠（或成摞、成堆）地放在…，（使）放成整齐的一叠
draught	n.	通风，穿堂风
circulate	v.	循环，（液体或气体）环流

Comprehension

1. Which of the following is right? _____.

   A. During the combustion of coal, a high temperature is produced inside the boiler. This temperature is high enough to convert water into steam.

   B. During the combustion of coal, a low temperature is produced inside the boiler. This temperature is high enough to convert water into steam.

C. During the combustion of coal, a low temperature is produced inside the boiler. This temperature is high enough to convert gas into steam.

D. During the combustion of coal, a low temperature is produced inside the boiler. This temperature is high enough to convert water into steam.

2. The boiler _____ in the thermal power plant is used to convert the chemical energy of coal into thermal energy or heat energy.

    A. boiler   B. turbine   C. condenser   D. economizer

3. The turbine is a mechanical device that is used to convert the _____ energy of steam into rotational energy or kinetic energy.

    A. nuclear   B. heat   C. wind   D. solar

4. The rotor of the alternator rotates and generates _____ energy.

    A. mechanical   B. electrical   C. kinetic   D. rotational

5. During the combustion of coal, the _____ are generated in the boiler.

    A. pressures   B. voltages   C. flue gases   D. currents

Unit 18　Solar Energy

Solar energy, radiation from the Sun is capable of producing heat, causing chemical reactions, or generating electricity. The total amount of solar energy incident on Earth is vastly in excess of the worlds current and anticipated energy requirements. If suitably harnessed, this highly diffused source has the potential to satisfy all future energy needs. In the 21st century, solar energy has become increasingly attractive as a renewable energy source because of its inexhaustible supply and its nonpolluting character, in stark contrast to the finite fossil fuel coal, petroleum, and natural gas.

Importance and potential

The Sun is an extremely powerful energy source, and sunlight is by far the largest source of energy received by Earth, but its intensity at earth's surface is actually quite low. This is essentially because of the enormous radial spreading of radiation from the distant Sun. A relatively minor additional loss is due to earth's atmosphere and clouds, which absorb or scatter as much as 54 percent of the incoming sunlight[1]. The sunlight that reaches the ground consists of nearly 50 percent visible light, 45 percent infrared radiation, and smaller amounts of ultraviolet and other forms of electromagnetic radiation.

Solar energy drives and affects countless natural processes on Earth. For example, photosynthesis by plants, algae and cyanobacteria relies on energy from the Sun, and it is nearly impossible to overstate the importance of that process in the maintenance of life on Earth. If photosynthesis ceased, there would soon be little food or other organic matter on Earth. Most organisms would disappear, and in time Earths atmosphere would become nearly devoid of gaseous oxygen. Solar energy is also essential for the evaporation of water in the water cycle, land and water temperatures, and the formation of wind, all of which are major factors in the climate patterns that shape life on Earth[2].

The potential for solar energy to be harnessed as solar power is enormous, since about 200,000 times the worlds total daily electric-generating capacity is received by Earth every day in the form of solar energy[3]. Unfortunately, though solar energy itself is free, the high cost of its collection, conversion, and storage still limits its exploitation in many places. Solar radiation can be converted either into thermal energy (heat) or into electrical energy, though the former is easier to accomplish.

Uses

Solar energy has long been used directly as a source of thermal energy. Beginning in the 20th century, technological advances have increased the number of uses and applications of the Suns thermal energy and opened the doors for the generation of solar power.

Thermal energy

Among the most common devices used to capture solar energy and convert it to thermal energy are flat-plate collectors, which are used for solar heating applications. Because the intensity of solar radiation at Earths surface is so low, these collectors must be large in area. Even in sunny parts of the worlds temperate regions, for instance, a collector must have a surface area of about 40 square meters (430 square feet) to gather enough energy to serve the energy needs of one person[4].

The most widely used flat-plate collectors consist of a blackened metal plate, covered with one or two sheets of glass, that is heated by the sunlight falling on it[5]. This heat is then transferred to air or water, called carrier fluids, which flow past the back of the plate. The heat may be used directly, or it may be transferred to another medium for storage. Flat-plate collectors are commonly used for solar water heaters and house heating. The storage of heat for use at night or on cloudy days is commonly accomplished by using insulated tanks to store the water heated during sunny periods. Such a system can supply a home with hot water drawn from the storage tank, or, with the warmed water flowing through tubes in floors and ceilings, it can provide space heating. Flat-plate collectors typically heat carrier fluids to temperatures ranging from 66 to 93 °C (150 to 200 °F). The efficiency of such collectors (i.e., the proportion of the energy received that they convert into usable energy) ranges from 20 to 80 percent, depending on the design of the collector.

Another method of thermal energy conversion is found in solar ponds, which are bodies of salt water designed to collect and store solar energy. The heat extracted from such ponds enables the production of chemicals、food、textiles and other industrial products and can also be used to warm greenhouses, swimming pools, and live stock buildings. Solar ponds are sometimes used to produce electricity through the use of the organic Rankine cycle engine, a relatively efficient and economical means of solar energy conversion, which is especially useful in remote locations. Solar ponds are fairly expensive to install and maintain and are generally limited to warm rural areas.

On a smaller scale, the Suns energy can also be harnessed to cook food in specially designed solar ovens. Solar ovens typically concentrate sunlight from over a wide area to a central point, where a black-surfaced vessel converts the sunlight into heat. The ovens are typically portable and require no other fuel inputs.

Electricity generation

Solar radiation may be converted directly into solar power (electricity) by solar cells(Fig. 4.3),or photovoltaic cells. In such cells, a small electric voltage is generated when light strikes the junction between a metal and a semiconductor (such as silicon) or the junction between two different semiconductors. (see photovoltaic effect.) Small photovoltaic cells that operate on sunlight or artificial light have found major use in low-power applications—for example, as power sources for calculators and watches. Larger units have been used to provide power for water pumps and communications systems in remote areas and for weather and communications satellites. By connecting large numbers of individual cells together, however, as in solar-panel arrays, hundreds or even thousands of kilowatts of electric power can be generated in a solar electric plant or in a large household array[6].

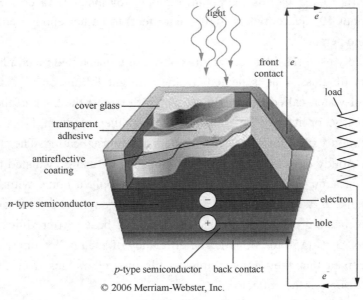

Fig.4.3　Solar cell

Concentrated solar power plants(Fig.4.4) employ concentrating, or focusing collectors to concentrate sunlight received from a wide area onto a small blackened receiver, thereby considerably increasing the light intensity in order to produce high temperatures. The arrays of carefully aligned mirrors or lenses can focus enough sunlight to heat a target to temperatures of

2000°C (3600°F) or more. This heat can then be used to operate a boiler, which in turn generates steam for a steam turbine electric generator power plant. For producing steam directly, the movable mirrors can be arranged so as to concentrate large amounts of solar radiation upon blackened pipes through which water is circulated and thereby heated.

Fig.4.4　Concentrated solar-power plant

Other applications

Solar energy is also used on a small scale for purposes other than those described above. In some countries, for instance, solar energy is used to produce salt from seawater by evaporation. Similarly, solar-powered desalination units transform salt water into drinking water by converting the Suns energy to heat, directly or indirectly, to drive the desalination process.

Solar technology has also emerged for the clean and renewable production of hydrogen as an alternative energy source. Mimicking the process of photosynthesis, artificial leaves are silicon-based devices that use solar energy to split water into hydrogen and oxygen, leaving virtually no pollutants. Further work is needed to improve the efficiency and cost-effectiveness of these devices for industrial use.

Technical Words and Expressions

solar energy			太阳能
chemical reaction			化学反应
harness	[ˈhɑːnɪs]		控制，利用
inexhaustible	[ˌɪnɪɡˈzɔːstəbl]	adj.	取之不尽，用之不竭的，无穷无尽的
fossil	[ˈfɒsl]	adj. & n.	化石的；化石
fossil fuel			化石燃料
petroleum	[pəˈtrəʊliəm]	n.	石油，原油
natural gas			天然气
minor	[ˈmaɪnə(r)]	adj.	少数的，轻微的，次要的，较小的
scatter	[ˈskætə(r)]	v.	散射
infrared	[ˌɪnfrəˈred]	adj. & n.	红外线的，使用红外线的；红外线，红外区
infrared radiation			红外辐射
ultraviolet	[ˌʌltrəˈvaɪələt]	adj. & n.	紫外线的，利用紫外线的；紫外线，紫外光，紫外线辐射

英文	音标	词性	中文
electromagnetic radiation			电磁辐射
photosynthesis	[ˌfəʊtəʊˈsɪnθəsɪs]	n.	光合作用
algae	[ˈældʒiː]	n.	藻类
cyanobacteria	[saɪænɒˈbæktɪərɪə]	n.	蓝藻
overstate	[ˌəʊvəˈsteɪt]	vt.	夸大，夸张，言过其实
organic	[ɔːˈgænɪk]	adj.&n.	有机的，器官的，生物的，有机物的；分子有机物
organic matter			有机物
organism	[ˈɔːgənɪzəm]	n.	生物，（尤指）微生物，有机体，有机组织，有机体系
devoid	[dɪˈvɔɪd]	adj.	缺乏，完全没有
oxygen	[ˈɒksɪdʒən]	n.	氧气，氧
gaseous oxygen			气态氧
exploitation	[ˌeksplɔɪˈteɪʃn]	n.	利用，开发，开采
flat-plate collector			平板收集器，平板组收集极，平板集能器，太阳能集热器
temperate region			温带
carrier	[ˈkærɪə(r)]	n.	载体，航空母舰，（尤指经营空运的）运输公司
carrier fluid			载液，携带液，载流体，载送流体
insulated tank			保温储罐
ceiling	[ˈsiːlɪŋ]	n.	天花板，上限，顶棚
solar pond			太阳池，太阳能池
extract	[ˈekstrækt, ɪkˈstrækt]	vt.	提取，获得，得到
textile	[ˈtekstaɪl]	n.	纺织品，纺织原料，纺织业
greenhouse	[ˈgriːnhaʊs]	n.	温室，暖房，温室效应
live stock			牲畜
organic Rankine cycle engine			有机朗肯循环发动机
rural area			农村地区
oven	[ʌvn]	n.	烤箱，烤炉
solar oven			太阳能炉
vessel	[ˈvesl]	n.	容器
solar cell			太阳能电池
photovoltaic	[ˌfəʊtəʊvalˈteɪɪk]	adj.	光伏的，光电池的
photovoltaic cell			光伏电池
artificial light			人工照明，人造光
solar-panel array			太阳能电池板阵列

Chapter 4　Power Supply Technology

lense	[lenz]	n.	透镜
desalination	[ˌdiːˌsælɪˈneɪʃn]	n.	海水淡化，（海水的）脱盐
desalination unit			海水淡化装置
hydrogen	[ˈhaɪdrədʒən]	n.	氢，氢气
alternative energy source			替代能源
mimick		vt.	模仿，模仿，摹拟
pollutant	[pəˈluːtənt]	n.	污染物，污染物质

 Notes

[1] A relatively minor additional loss is due to earth's atmosphere and clouds, which absorb or scatter as much as 54 percent of the incoming sunlight.

译文：相对较小的额外损失是由于地球的大气层和云层，它们吸收或散射了多达 54%的入射阳光。

注释：which 引导非限制性定语从句，which 是指 "earth's atmosphere and clouds" 在从句中作主语。

[2] Solar energy is also essential for the evaporation of water in the water cycle, land and water temperatures, and the formation of wind, all of which are major factors in the climate patterns that shape life on Earth.

译文：太阳能对水循环中的水分蒸发、土地和水温以及风的形成也至关重要，所有这些都是塑造地球上生命的气候模式的主要因素。

注释："all of which"引导的是非限制性定语从句，which 指代的是"water in the water cycle, land and water temperatures, and the formation of wind，"所以从句中谓语用复数形式 are。

[3] The potential for solar energy to be harnessed as solar power is enormous, since about 200,000 times the worlds total daily electric-generating capacity is received by Earth every day in the form of solar energy.

译文：太阳能被用作太阳能的潜力是巨大的，因为地球每天以太阳能的形式接收的发电量约为世界日总发电量的 20 万倍。

注释："to be harnessed as solar power"为不定式短语做后置定语，句中 since 引导原因状语从句。

[4] Even in sunny parts of the worlds temperate regions, for instance, a collector must have a surface area of about 40 square meters (430 square feet) to gather enough energy to serve the energy needs of one person.

译文：例如，即使在世界温带阳光充足的地区，收集器的表面积也必须约为 40 平方米（430 平方英尺），才能收集到足够的能量来满足一个人的能源需求。

注释："for instance"为插入语，用在句中。"to gather enough energy"用作目的状语。

[5] The most widely used flat-plate collectors consist of a blackened metal plate, covered with one or two sheets of glass, which is heated by the sunlight falling on it.

译文：最广泛使用的平板收集器由一块发黑的金属板组成，上面覆盖着一两片玻璃，被落在上面的阳光加热。

注释:"covered with one or two sheets of glass"为过去分词作定语修饰"a blackened metal plate","falling on it"现在分词短语作"sunlight"的后置定语。

[6] By connecting large numbers of individual cells together, however, as in solar-panel arrays, hundreds or even thousands of kilowatts of electric power can be generated in a solar electric plant or in a large household array.

译文:然而,通过将大量的单个电池连接在一起,就像在太阳能电池板阵列中一样,可以在太阳能发电厂或大型家庭阵列中产生数百甚至数千千瓦的电力。

注释:"as in solar-panel arrays" as 引导方式状语,表示"以…方式""像…一样"。

主句的主语为"hundreds or even thousands of kilowatts of electric power",By 引导方式状语。

Exercises

I. Mark the following statements with T(true) or F(false) according to the text.

1. The total amount of solar energy incident on Earth is vastly in excess of the worlds current and anticipated energy requirements. (　　)

2. The Sun is an extremely powerful energy source, and sunlight is by far the largest source of energy received by Earth, even its intensity at earth's surface is quite large. (　　)

3. Fortunately, since solar energy itself is free, the low cost of its collection, conversion, and storage opens its exploitation in many places. (　　)

4. Flat-plate collectors typically heat carrier gases to temperatures ranging from 66 to 93 °C (150 to 200 °F). (　　)

5. Concentrated solar power plants employ concentrating, or focusing collectors to concentrate sunlight received from a wide area onto a small blackened receiver, thereby considerably increasing the lights intensity in order to produce high temperatures. (　　)

II. Complete the following sentences.

1. In the 21st century solar energy has become increasingly attractive as a renewable energy source because of its _____ supply and its _____ character, in stark contrast to the finite fossil fuels: coal, petroleum, and natural gas.

2. The sunlight that reaches the ground consists of nearly 50 percent _____, 45 percent _____, and smaller amounts of _____ and other forms of electromagnetic radiation.

3. Unfortunately, though solar energy itself is free, the high _____ of its collection, conversion, and storage still limits its _____ in many places.

4. The most widely used flat-plate collectors consist of a blackened _____ plate, covered with one or two sheets of _____, that is heated by the _____ falling on it.

5. Solar radiation may be converted directly into solar power (electricity) by _____ Or _____ cells.

Reading 18 What is Wind Energy

Wind energy is electricity created from the naturally flowing air in the Earth's atmosphere. As a renewable resource that won't get depleted through use, its impact on the environment and climate crisis is significantly smaller than burning fossil fuels.

Wind energy can be created by something as simple as a set of 8-foot sails positioned to capture prevailing winds that then turn a stone and grind grain (a gristmill). Or it can be as complex as a 150-foot vane turning a generator that produces electricity to be stored in a battery or deployed over a power distribution system. There are even bladeless wind turbines.

As of 2021, there are over 67,000 wind turbines running in the United States, found in 44 states, Guam, and Puerto Rico. Wind generated about 8.4% of the electricity in the U.S. in 2020. Worldwide, it provides about 6% of the world's electricity needs. Wind energy is growing year-over-year by about 10% and is a key part of most climate change reduction and sustainable growth plans in a variety of countries, including China, India, Germany, and the United States.

Wind energy definition

Human beings use wind energy in a variety of ways, from the simple (it's still used to pump water for livestock in more remote locations) to the increasingly complex—think of the thousands of turbines that dominate the hills that cut through highway 580 in California (Fig.4.5).

Fig.4.5 Rows of wind turbine

The basic components of any wind energy system are fairly similar. There are blades of some size and shape that are connected to a drive shaft, and then a pump or generator that either uses or collects the wind energy. If the wind energy is used directly as a mechanical force, like milling grain or pumping water, it's called a windmill; if it converts wind energy to electricity, it's known as a wind turbine. A turbine system requires additional components, such as a battery for electricity storage, or it may be connected to a power distribution system like power lines.

Nobody really knows when the wind was first harnessed by a human being, but wind was definitely being utilized as a way to move boats on Egypt's Nile River around the year 5,000 BC By 200 BC people in China were using the wind to power simple water pumps, and windmills with hand-woven blades were used to grind grain in the Middle East. Over time, wind pumps and mills were used in all kinds of food production there, and the concept then spread to Europe, where the Dutch built large wind pumps to drain wetlands and from there the idea traveled to the Americas.

Wind energy basics

Wind is produced naturally when the sun heats the atmosphere, from variations in the surface of the Earth, and from the planet's rotation. Wind can then increase or decrease as a result of the influence of bodies of water, forests, meadows and other vegetation, and elevation changes. Wind patterns and speeds vary significantly across terrain, as well as seasonally, but some of those patterns are predictable enough to plan around.

Site selection

The best locations to place a wind turbine are the tops of rounded hills, on open plains (or open water for offshore wind), and mountain passes where wind is naturally funneled through (producing regular high wind speeds). Generally, the higher the elevation the better, since higher elevations usually have more wind.

Wind energy forecasting is an important tool for siting a wind turbine. There are a variety of wind speed maps and data from the National Oceanic and Atmospheric Administration (NOAA) or the National Renewable Energy Laboratory (NREL) in the U.S. that provide these details.

Then, a site-specific survey should be done to assess the local wind conditions and to determine the best direction to place the wind turbines for maximum efficiency. For at least a year, projects on land track wind speed, turbulence, and direction, as well as air temperatures and humidity. Once that information is determined, turbines that will deliver predictable results can be built.

Wind isn't the only factor for siting turbines. Developers for a wind farm must consider how close the farm is to transmission lines (and cities that can utilize the power); possible interference to local airports and plane traffic; underlying rock and faults; flight patterns of birds and bats; and local community impact (noise and other possible effects).

Most larger wind projects are designed to last at least 20 years, if not more, so these factors must be considered over the long term.

Types of wind energy

Utility scale wind energy

These are large-scale wind projects designed to be used as a source of energy for a utility company. They are similar in scope to a coal-fired or natural gas power plant, which they sometimes replace or supplement. Turbines exceed 100 kilowatts of power in size and are usually installed in groups to provide significant power currently these types of systems provide about 8.4% of all energy in the United States.

Offshore wind energy

These are generally utility-scale wind energy projects that are planned in the waters off coastal areas. They can generate tremendous power near larger cities (which tend to cluster closer to shore in much of the United States). Wind blows more consistently and strongly in offshore areas than in land, according to the U.S. Department of Energy. Based on the organization's data and calculations, the potential for offshore wind energy in the U.S. is more than 2000 gigawatts of power, which is two times the generating capacity of all U.S. electric power plants. Worldwide, wind energy could provide more than 18 times what the world currently uses, according to the International Energy Agency.

Small scale or distributed wind energy

This type of wind energy is the opposite of the examples above. These are wind turbines that are smaller in physical size and are used to meet the energy demands of a specific site or local area. Sometimes, these turbines are connected to the larger energy distribution grid, and sometimes they are off-grid. You'll see these smaller installations (5 kilowatt size) in residential settings, where they might provide some or most of a home's needs, depending on weather, and medium-sized versions (20 kilowatts or so) at industrial or community sites, where they might be part of a renewable energy system that also includes solar power, geothermal, or other energy sources.

How does wind energy work?

The function of a wind turbine is to use blades of some shape (which can vary) to catch the wind's kinetic energy. As the wind flows over the blades, it lifts them, just like it lifts a sail to push a boat. That push from the wind makes the blades turn, moving the drive shaft that they're connected to. That shaft then turns a pump of some kind—whether directly moving a piece of stone over grain (windmill), or pushing that energy into a generator that creates electricity that can be used right away or stored in a battery.

The process for an electricity-generating system (wind turbine) includes the following steps.

Wind pushes blades

Ideally, a windmill or wind turbine is located in a place with regular and consistent winds. That air movement pushes specially designed blades that allow the wind to push them as easily as possible. Blades can be designed so they are pushed upwind or downwind of their location.

Kinetic energy is transformed

Kinetic energy is the free energy that comes from the wind. For us to be able to use or store that energy, it needs to be changed into a usable form of power. Kinetic energy is transformed into mechanical energy when the wind meets the windmill blades and pushes them. The movement of the blades then turns a drive shaft.

Electricity is generated

In a wind turbine, a spinning drive shaft is connected to a gearbox that increases the speed of the rotation by a factor of 100—which in turn spins a generator. Therefore, the gears end up spinning much faster than the blades being pushed by the wind. Once these gears reach a fast enough speed, they can power a generator that produces electricity.

The gearbox is the most expensive and heavy part of the turbine, and engineers are working on direct drive generators that can operate at lower speeds (so they don't need a gear box).

Transformer converts electricity

The electricity produced by the generator is 60-cycle AC (alternating current) electricity. A transformer may be needed to convert that to another type of electricity, depending on local needs.

Electricity is used or stored

Electricity produced by a wind turbine might be used on site (more likely to be true in small or medium-sized wind projects), it could be delivered to transmission lines for use right away, or it could be stored in a battery.

More efficient battery storage is key for advancements in wind energy in the future. Increased storage capacity means that on days when the wind blows less, stored electricity from windier days could supplement it. Wind variability would then become less of an obstacle for reliable electricity from wind.

What is a wind farm?

A wind farm is a collection of wind turbines that form a type of power plant, producing electricity from wind. There's no official number requirement for an installation to be considered a wind farm, so it could include a few or hundreds of wind turbines working in the same area, whether on land or offshore.

Technical Words and Expressions

wind energy		风能
climate crisis		气候危机
prevailing	adj.	普遍的，盛行的，流行的，（指风）一地区常刮的
vane	n.	（风车等的）翼，叶片，轮叶
deploy	vt.	部署
power distribution system		配电系统，电源分配系统
wind turbine		风力涡轮机，风力发电机
sustainable	adj.	可持续的
windmill		风车
grind	v.	磨碎
mill	n.	磨粉机
meadow	n.	草地，牧场

vegetation	n.	植被，（统称）植物，草木
elevation	n.	（尤指）海拔，高处，高地
wind pattern		风向
terrain	n.	地形，地势
offshore	adj.	海上的，离岸的，近海的
funnel	vt.	（使）流经狭窄空间，经过漏斗形口子
assess	vt.	评估；评定
turbulence	n.	湍流；紊流；动荡；（空气和水的）涡流
coastal	adj.	沿海的，靠近海岸的
utility	n.	实用，实用程序，效用，公用事业
grid	n.	（利用互联网的）联网，电网
off-grid	n.	离网
geothermal	adj.	地热的
gearbox	n.	齿轮箱，变速箱
transformer	n.	变压器
obstacle	n.	障碍，障碍物，阻碍，绊脚石

Comprehension

1. Wind energy is growing year-over-year by about _____ % and is a key part of most climate change reduction and sustainable growth plans in a variety of countries, including China, India, Germany, and the United States.

   A. 5   B. 10   C. 15   D. 20

2. The best locations to place a wind turbine are _____.

   A. the tops of rounded hills

   B. on open plains

   C. mountain passes where wind is naturally funneled through

   D. all above

3. The function of a wind turbine is to use blades of some shape (which can vary) to catch the wind's _____.

   A. static energy   B. kinetic energy

   C. electrical energy   D. potential energy

4. A transformer may be needed to convert _____ to _____, depending on local needs.

   A. ac ac   B. ac dc   C. dc dc   D. dc ac

5. The _____ is the most expensive and heavy part of the turbine, and engineers are working on direct drive generators that can operate at lower speeds.

   A. blade   B. gearbox   C. transformer   D. mill

Unit 19 Transmission and Distribution of Electricity

The place where electric power is produced by the parallel connected three phase alternators/generators is called Generating Station (i.e. power plant).

The ordinary power plant capacity and generating voltage may be 11kV, 11.5kV, 12kV or 13kV. But economically, it is good to step up the produced voltage from (11kV, 11.5kV or 12kV) to 132kV, 220kV or 500kV or more (in some countries, up to 1500kV) by step up transformer (power Transformer).

Generation is the part of power system where we convert some form of energy into electrical energy. This is the source of energy in the power system. It keeps running all the time. It generates power at different voltage and power levels depending upon the type of station and the generators used. The maximum number of generators generate the power at voltage level around 11~20kV. The increased voltage level leads to the greater size of generator required and hence the cost involved[1].

Primary transmission

The electric supply (in 132kV, 220kV, 500kV or greater) is transmitted to the load center by three phase three wire (3-Phase 3-Wire also known as Delta connection) overhead transmission system.

As the voltage level which is generated is around 11~20kV, the demand is at various levels of voltage and the load center is at very far away places from the generating station, For example, the generating station can be generating voltage at 11kV, but the load center is 1000km apart and at the level of 440V. Therefore, for the delivery of electrical energy at such a long distance, an arrangement must be there to make it possible. Hence, the transmission system is essential for the delivery of electrical energy. This is made possible by using transmission lines of different length. These are overhead transmission lines in almost every case. Some exceptions occur when it is needed to cross an ocean, then there is a compulsion to use underground cables.

But, as the system grew and load demand increased, the challenge in this process has become very complex. At low voltage level, the amount of current flowing through the line for high load demand is more and hence the voltage drop due to the resistance and reactance of the transmission line is very significant. This leads to more losses in the transmission lines and a decrease in the voltage at the load end.

This affects the cost of the system and the working of the equipment the consumers use. So, transformer is used to increase the voltage level at certain values ranging from 220kV to 765kV. This makes the current value lesser for the same load that would be having higher values of current at certain load.

The increased demand and the constraint of location of generating station have made possible the need of a very complex system called "Grid". This system connects multiple generating stations generating voltage at different levels being connected together as a combined system.

This makes the system to reach out to various load centers and this provides a great system of having higher reliability. Presently, this system has grown to size of a country. One more system being used nowadays is the use of HVDC. HVDC is used for greater distances and sometimes used to connect two grids of different voltage or frequency levels. HVDC also provides lower corona losses, lower communication interferences, elimination of inductive effect and elimination of frequency of operation.

Transmission lines vary in sizes. This size determines its characteristics and its behavior in the system. For example, in long transmission lines the voltage at the consumer end becomes higher than its rated value during light load condition due to the dominating capacitive nature of the transmission lines.

Secondary transmission

In area far from the city (outskirts), the transmission which has been connected with receiving stations by lines is called secondary transmission. At the receiving station, the level of voltage is reduced by step down transformers to 132kV, 66 or 33kV. And electric power is transferred by three phase three wire(3-Phase 3-Wire) overhead system to different substations.

Primary distribution

At a substation, the level of secondary transmission voltage (132kV, 66 or 33kV) is reduced to 11kV by step-down transforms.

Generally, electric supply is provided to those heavy load consumer (commercial power supply for industries) where the demands is 11kV, from the lines which caries 11kV (in three phase three wire overhead system) and they make a separate substation to control and utilize the heavy power in industries and factories.

In other cases for heavier load consumers (at large scale), the demand is up to132kV or 33 kV. So electric supply is provided directly by secondary transmission or primary distribution (in 132kV, 66kV or 33kV) and, then stepped down by step-down transformers in their own substation for utilization (i.e. for electric traction etc).

When the transmission lines get closer to the demand centers, the voltage level is reduced to make it practical to distribute at different places of load. Therefore, power is taken from the grid and stepped down to 30~33kV, depending upon the places where it is being delivered[2]. This is then transmitted to substations. For example, the system voltage at substation level in India is 33kV.

Many control mechanisms are provided in the substations to make the power delivery a controlled and continuous process without much disturbance. These substations deliver power to smaller units called 'Feeders'. This is done by either 'Overhead lines' or 'Underground cables'. These feeders are in towns, cities, or villages or it may be some group of industries, which takes the power from the substation, and converts its voltage level according to its own use.

For domestic use, the voltage is further reduced at 110~230V (phase to ground) to be used by the individuals at different power factor[3]. The combined amount of demand is the load on the entire system and that must be generated at that instant.

Depending upon the scheme of the distribution system, it is categorized as radial or ring mains. It gives a different degree of reliability and stability to the system. All these systems are protected using various protection schemes comprising of circuit breakers, relays, lightening arresters, ground wires etc[4].

Many measuring and sensing elements are also associated like "current transformer" and "potential transformer" and metering at all the places from the substations to feeders to the consumers places.

Secondary distribution

Electric power is transferred (from primary distribution line i.e.11kV) to a distribution substation known as secondary distribution. This substation is located nearby domestic & consumers areas where the level of voltage is reduced to 440V by step-down transformers.

These transformers are called distribution transformers with three-phase four-wire system (3-Phase 4-Wire also known as Star connection). So there is 400 Volts (Three Phase Supply System) between any two phases and 230 Volts (Single Phase Supply) between a neutral and phase (live) wires.

Residential load (i.e. Fans, Lights, and TV etc) may be connected between any one phase and neutral wires, while three phase load may be connected directly to the three phase lines.

In short, secondary power distribution may be divided into three sections such as feeders, distributors and service lines .

🔌 Technical Words and Expressions

three phase alternator			三相交流发电机
step up			增加（数量），提高（速度、强度等）
three-phase three-wire			三相三线制
compulsion	[kəmˈpʌlʃn]	n.	强迫，强制
line voltage			线电压
corona	[kəˈrəʊnə]	n.	（尤指在日食或月食期间的）日冕，日华
underground cable			地下电缆
step down			降压
overhead line			架空线路，架空线
categorize	[ˈkætəgəraɪz]	vt.	分类，将...分类；把...加以归类
ring main			环形干线
circuit breaker			断路器，<美>断路开关
arrester	[əˈrɛstə]	n.	避雷器，制动装置

lightening arrester			避雷器，避雷装置
ground wire			地线
substation	['sʌbsteɪʃn]	n.	变电站，变电所
domestic	[də'mestɪk]	adj.	国内的，家庭的，本国的
distribution transformer			配电变压器
star connection			星形连接，星形接线，星形连接，星形连接
neutral	['nju:trəl]	adj.	中性的
feeder	['fi:də(r)]	n.	支线，馈线
distributor	[dɪ'strɪbjətə(r)]	n.	（发动机的）配电器，配电盘
service line			动力管线

Notes

[1] The increased voltage level leads to the greater size of generator required and hence the cost involved.

译文：增加的电压等级导致所需的发电机的尺寸更大，从而导致所涉及的成本更高。

注释：required 作 the greater size of generator 的后置定语，involved 作 the cost 的后置定语。"the greater size of generator required" 和 "hence the cost involved" 作 leads to 的并列宾语。

[2] Therefore, power is taken from the grid and stepped down to 30~33kV, depending upon the places where it is being delivered.

译文：因此，根据输电地点的不同，电力从电网中提取并降压至 30~33kV。

注释："depending upon ..." 现在分词作方式状语。

[3] For domestic use, the voltage is further reduced at 110~230V (phase to ground) to be used by the individuals at different power factor.

译文：对于家用，电压进一步降低到 110~230V（相对地），供不同功率因数的个人使用。

注释："to be used by the individuals at different power factor." 作目的状语。

[4] All these systems are protected using various protection schemes comprising of circuit breakers, relays, lightening arresters, ground wires etc.

译文：所有这些系统都使用各种保护方案进行保护，包括断路器、继电器、避雷器、地线等。

注释："using ..." 作方式状语，修饰谓语动词 "protected"。"comprising of circuit breakers, relays, lightening arresters, ground wires etc" 作 "various protection schemes" 的后置定语。

Exercises

Ⅰ. Mark the following statements with T(true) or F(false) according to the text.

1. The place where electric power produced by the series connected 3-phase alternators/generators is called Generating Station (i.e. power plant).　　　　（　　）

2. The minimum number of generators generate the power at voltage level around 11~

20kV. ( )

3. These are overhead transmission lines in almost every cases. Some exceptions occur when it is needed to cross an ocean. ( )

4. HVDC is used for greater distances and sometimes used to connect two grids of different current or frequency levels. ( )

5. In short, secondary power distribution may be divided in three sections such as feeders, distributors and service lines . ( )

Ⅱ. Complete the following sentences.

1. Hence, the transmission system is essential for the _____ of electrical energy. This is made possible by using the _____ of different length.

2. HVDC also provides lower _____, lower _____, elimination of _____ and elimination of _____.

3. These substations deliver power to smaller units called 'Feeders'. This is done by either '_____' or '_____'.

4. Depending upon the scheme of the distribution system, it is categorized as_radial or _____.

5. In short, secondary power distribution may be divided in three sections such as feeders, _____ and _____.

Reading 19 What is Power Factor

In electrical engineering, the power factor (PF) of an AC electrical power system is defined as the ratio of working power (measured in kilowatts, kW) absorbed by the load to the apparent power (measured in kilovolt amperes, kVA) flowing through the circuit. Power factor is a dimensionless number in the closed interval of −1 to 1.

The "ideal" power factor is one (also referred to as "unity"). This is when there is no reactive power through the circuit, and hence the apparent power (kVA) is equal to the real power (kW). A load with a power factor of 1 is the most efficient loading of the supply.

That said this is not realistic, and the power factor will in practice be less than 1. Various power factor correction techniques are used to help increase the power factor to this ideal state.

In circuits with resistors, inductors, and capacitors, a phase difference naturally occurs between the source voltage and the current.

The cosine of this phase difference is called the electrical power factor. This factor ($-1 < \cos\varphi < 1$) represents the fraction of the total power that is used to do the useful work.

Power factor improvement

The term power factor comes into the picture in AC circuits only. Mathematically it is the cosine of the phase difference between the source voltage and current. It refers to the fraction of total power (apparent power) which is utilized to do the useful work called active power.

The needs for power factor improvement are shown below.
- Real power is given by $P = UI\cos\varphi$. The electrical current is inversely proportional to $\cos\varphi$ for transferring a given amount of power at a certain voltage. Hence higher the PF, lower will be the current flowing. A small current flow requires a less cross-sectional area of conductors, and thus it saves conductors and money.
- From the above relation, we see having a poor power factor increases the current flowing in a conductor, and thus copper loss increases. A large voltage drop occurs in the alternator, electrical transformer, and transmission, and distribution lines which gives very poor voltage regulation.
- The kVA rating of machines is also reduced by having a higher power factor, as per the formula:

$$\cos\varphi = \frac{Active\ power}{Apparent\ power}$$

Hence, the size and cost of the machine are also reduced.

This is why the electrical power factor should be maintained close to unity it is significantly cheaper.

Methods of power factor improvement

There are three main ways to improve power factor:
- Capacitor banks.
- Synchronous condensers.
- Phase advancers.

Capacitor banks

Improving power factor means reducing the phase difference between voltage and current. Since the majority of loads are of inductive nature, they require some amount of reactive power for them to function.

A capacitor or bank of capacitors installed parallel to the load provides this reactive power. They act as a source of local reactive power, and thus less reactive power flows through the line.

Capacitor banks reduce the phase difference between the voltage and current.

Synchronous condensers

Synchronous condensers are 3-phase synchronous motors with no load attached to their shaft.

The synchronous motor has the characteristics of operating under any power factor leading, lagging, or unity depending upon the excitation. For inductive loads, a synchronous condenser is connected towards the load side and is overexcited.

Synchronous condensers make it behave like a capacitor. It draws the lagging current from the supply or supplies the reactive power.

Phase advancers

This is an AC exciter mainly used to improve the PF of an induction motor.

Mounted on the motors shaft and connected to its rotor circuit, phase advancers enhance the power factor by supplying the necessary ampere turns for the required magnetic flux at a specific slip frequency.

Further, if ampere-turns increase, it can be made to operate at the leading power factor.

Technical Words and Expressions

apparent power		视在功率
dimensionless number		无量纲数
reactive power		无功功率
copper loss		铜损耗
capacitor bank		电容器组
phase difference		相位差
synchronous condenser		同步调相机，调相机，同步电容器，同步进相机
overexcite	vt.	过励
exciter	n.	励磁机
phase advancer		相位超前补偿器

Comprehension

1. In electrical engineering, the power factor (PF) of an AC electrical power system is defined as the ratio of _____ absorbed by the load to the _____ flowing through the circuit.
    A. apparent power，reactive power
    B. working power，reactive power
    C. working power，apparent power
    D. reactive power，working power

2. In circuits with resistors, inductors, and capacitors, a phase difference naturally occurs between the _____ and the _____.
    A. source voltage，current
    B. source voltage，power
    C. current，power
    D. current，charge

3. The _____ of this phase difference is called the electrical power factor.
    A. sine
    B. cosine
    C. tangent
    D. Cotangent

4. A capacitor or bank of capacitors installed _____ to the load provides this _____ power.
    A. series,reactive
    B. series,apparent
    C. parallel,reactive
    D. parallel, apparent

5. Further, if ampere-turns _____, it can be made to operate at the _____ power factor.
    A. increase,lagging
    B. increase,leading
    C. decrease,lagging
    D. decrease,leading

Unit 20 Difference between AC and DC Transmission System

Electric power can be transmitted in both AC and DC for short and long transmission and distribution systems.

There are some advantages and disadvantages of both systems. Lets discuss the technical advantages and disadvantages of both AC and DC power transmission lines systems.

DC transmission

In earlier times, the electric power transmission was done in DC due to the following advantages over AC.

Advantages of DC transmission
- There are two conductors used in DC transmission, while three conductors are required in AC transmission[1].
- There are no inductance and surges (High Voltage waves for a very short time) in DC transmission.
- Due to the absence of inductance, there are very low voltage drop in DC transmission lines as compared to the AC (if both load and sending end voltage are the same)[2].
- There is no concept of skin effect in DC transmission lines. Therefore, a conductor having a small cross sectional area is required in the DC transmission line.
- A DC system has less potential stress over an AC system for the same Voltage level. Therefore, a DC line requires less insulation.
- In the DC system, there is no interference with other communication lines and systems.
- In the DC line, corona losses are very low as compared to the AC transmission lines.
- In High voltage DC (HVDC) transmission lines, there are no dielectric losses.
- In the DC transmission system, there are no difficulties in synchronizing and related stability problems.
- The DC system is more efficient than AC. The rate of price of towers, poles, insulators, and conductors are low so the system is economical[3].
- In DC system, the speed control range is greater than the AC system.
- There is low insulation needed in the DC system (about 70%).
- The price of DC cables is low (due to low insulation).
- In the DC supply system, the sheath losses in underground cables are low.
- The DC system is suitable for high power transmission based on high current transmission.
- In the DC system, the value of charging current is quite low. Therefore, the length of the DC transmission lines is greater than AC lines.

Disadvantages of DC transmission:
- Due to commutation problems, electric power can't be produced at High (DC)Voltage.
- In high voltage transmission, we can't step up the level of DC voltage (As transformers wont work on DC).
- There is a limitation of DC switches and circuit breakers (and they are costly too).
- The motor generator set is used to step down the level of DC voltage and the efficiency of the motor-generator set is lower than a transformer.
- The DC transmission system is more complex and costly as compared to the AC transmission system..
- The level of DC Voltage can not be changed (step-up or step-down) easily. So we can not get the desired voltage for electrical and electronics appliances (such as 5 Volts, 9 Volts ,15 Volts, 20 and 22 Volts etc) directly from the transmission and distribution lines.

AC transmission

Nowadays, the generation, transmission and distribution of electric power has mostly been in AC.
Advantages of AC transmission system
- AC circuit breakers are cheaper than DC circuit breakers.
- The repairing and maintenance of the AC substation is easier and inexpensive than the DC substation.
- The level of AC voltage may be increased or decreased by using step-up and step-down transformers.

Disadvantages of AC system
- In the AC line, the size of the conductor is greater than the DC Line.
- The cost of AC transmission lines is greater than DC transmission lines.
- Due to the skin effect, the losses in the AC system are more.
- Due to the capacitance in AC transmission lines, a continuous power loss occurs when there is no load on the power lines or the line is open at all.
- There are some additional line losses due to inductance.
- More insulation is required in the AC transmission system.
- The corona losses occur in an AC transmission line system.
- AC transmission lines interfere with other communication lines.
- There are stability and synchronizing problems in the AC system.
- The AC transmission system is less efficient than the DC transmission system.
- There are difficulties in controlling the reactive power.

The above comparison shows that DC transmission system is better than the AC transmission system, but still the majority of power transmission is done in AC power lines due to cost and uses of transformers for changing the level of voltages at different levels for different purposes[4].

Although mercury arc rectifiers, thyratrons, diodes and semiconductors can be used to easily convert AC into DC and DC into AC. Therefore, some countries transmit the electric power through DC power lines. The range of these DC power transmissions is up to 100kV to 800kV+.

Technical Words and Expressions

skin effect			集肤效应
potential stress			电压应力
dielectric loss			电晕损耗
tower	['taʊə(r)]	n.	输电线路塔
pole	[pəʊl]	n.	电杆
sheath loss			护套损耗
mercury arc rectifier			水银弧整流器
thyratron	['θaɪrətrɑn]	n.	闸流管

Notes

[1] There are two conductors used in DC transmission, while three conductors are required in AC transmission.

译文：直流输电使用两条导线，而交流输电需要三条导线。

注释：本句中"while"作并列连词用，意思为"而，然后"，表示对比。句中"used in DC transmission"为过去分词作后置定语修饰"conductors"

[2] Due to the absence of inductance, there are very low voltage drop in DC transmission lines as compared to the AC (if both load and sending end voltage is same).

译文为：由于没有电感，与交流相比，直流输电线路的电压降非常低（如果负载和发送端电压相同）。

注释：本句的核心结构为there be 结构，表示存在某物或某人。句首的due to 是一个介词短语，放在句子的开头或中间，后面跟名词或名词性短语，表示原因或起因。

[3] The DC system is more efficient than AC. The rate of price of towers, poles, insulators, and conductors are low so the system is economical.

译文：直流输电系统比交流输电系统更高效，因此塔架、电杆、绝缘子和导线的价格较低，从而导致直流输电系统成本较低。

注释：该从句中又包含了一个"so"引导的结果状语从句"so the system is economical"。

[4] The above comparison shows that DC transmission system is better than the AC transmission system, but still the majority of power transmission is done in AC power lines due to cost and uses of transformers for changing the level of voltages at different levels for different purposes.

译文：上述比较表明，直流输电系统比交流输电系统更好，但为了降低成本和在不同应用场合能够使用变压器来改变电压值的大小，大部分电力传输仍采用交流输电系统。

注释："but still"在句中连接两个句子，意思上有转折之意。前一句主语部分为"The above

comparison", 谓语为 "shows", "that DC transmission system is better than the AC transmission system" 为宾语从句, 后一句主语为 "the majority of power transmission", 谓语部分为 "is done"。"due to" 是一个固定的短语, 意为 "因为, 由于", 常用来表示原因。

Exercises

Ⅰ. Mark the following statements with T(true) or F(false) according to the text.

1. There are three conductors used in DC transmission. ( )
2. There are inductance and surges in DC transmission. ( )
3. In the DC Transmission system, there are no difficulties in synchronizing and related stability problems. ( )
4. In DC system, the speed control range is greater than AC system. ( )
5. The AC transmission system is less efficient than the DC transmission system. ( )

Ⅱ. Complete the following sentences

1. _____ can be transmitted in both AC and DC for short and long transmission and distribution systems.
2. In the DC system, there is no _____ with other communication lines and systems.
3. The DC system is more efficient than AC, therefore, the rate of price of _____, _____, _____, and _____ are low so the system is economical.
4. The level of AC _____ may be increased or decreased by using step up and step down _____.
5. Due to the _____ in AC transmission lines, a continuous _____ occurs when there is no _____ on the power lines or line is open at all.

Reading 20 What are the Advantages of HVDC over HVAC

The electricity we consume travels very long distances before reaching us. The power generating stations, often in remote areas supply power that travels through several hundred miles & through multiple substations. The power transmission is done using very high voltage in order to reduce line losses. But the power can be in either AC or DC form and both modes of power transmission are being used. We are more accustomed to AC transmission because of the AC utility poles outside our homes & the available AC supply outlets in our home, but HVDC plays a vital role in power transmission with various advantages over HVAC.

The main objective in power transmission is to reduce the transmission power losses & supply power economically at a minimum possible investment. Both types of transmission are affected by several factors, but the HVDC transmission has more advantages than its disadvantages. In this article, we are going to discuss several such advantages of HVDC over HVAC transmission.

Low cost of transmission

The cost of transmission depends on various factors like the cost of equipment used for voltage conversion at the terminal stations, the number & size of conductors being used, transmission tower size and the power losses in the transmission, etc.

The equipment used for HVAC voltage conversion at the AC terminal station is mainly a transformer which is simpler & cheaper than HVDCs thyristors based converters. It is the only feature of HVAC transmission that surpasses the HVDC in terms of a minimum cost requirement.

The HVAC transmission requires a minimum of 3 conductors for 3-phase power transmission, while in case of HVDC that can utilize the earth as the return path can use only 1 conductor for mono-polar transmission or 2 conductors for a bipolar transmission. It substantially decreases the overall cost of the transmission. Even so, the 3 conductors used for 3-phase supply can be used for HVDC transmission with the capability to transmit double the same amount of power using a double bipolar link.

The HVAC transmission lines require relatively larger spacing between the phase to ground and the phase to phase conductors. In order to maintain such spacing, the transmission tower used for HVAC needs to be taller & wider than HVDC. Using the HVDC transmission tower reduces the installation cost as compared to HVAC towers.

The power losses in the HVDC transmission are significantly lower than HVAC. Therefore, HVDC is a more efficient form of transmission than HVAC.

The overall transmission cost can be broken down into two main categories i.e. Terminal station cost & transmission line cost. The former is a constant figure that does not depend on the distance of transmission while the latter depends on the distance of the transmission line. The terminal cost for AC is quite low while HVDC is very high. But the transmission line cost per 100 km for HVAC is far larger than that of the HVDC transmission line. Therefore, the overall cost graph for HVAC & HVDC meets at a point called break-even distance (Fig.4.6).

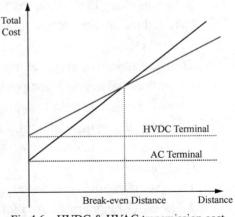

Fig.4.6 HVDC & HVAC transmission cost

The transmission distance at which the overall investment cost for HVAC starts increasing than HVDC is called break-even distance. This distance depends on the type of transmission. The break-even distance for overhead transmission is estimated at around 400~500m (600~800km) while the underwater transmission is 20~50km & underground is 50~100km. Therefore, the HVDC is a far more efficient & economically cheaper choice for power transmission over the break-even distance.

Reduced power losses

The HVDC transmission experiences very low losses as compared to HVAC. Here are some of the losses that are either completely eliminated or reduced significantly in HVDC.

Technical Words and Expressions

outlet	n.	[电]电源插座
HVDC		高压直流输电
HVAC		高压交流输电
terminal station		终端站，终点站
thyristor	n.	晶闸管
surpass	v.	超过，超越，优于，胜过
spacing	n.	（空间）间隔，尤指）字距，行距
break-even distance		盈亏平衡距离

Comprehension

1. Why we use very high voltage to transmit power? _____.
   A. because very high voltage is safer.
   B. because very high voltage is more efficient.
   C. to reduce the line losses
   D. all above

2. The main objective in power transmission is to _____.
   A. increase the transmission power losses
   B. increase the line losses
   C. reduce the transmission power losses and supply power economically at minimum possible investment
   D. all above

3. The cost of transmission depends on _____.
   A. the cost of equipment used for voltage conversion at the terminal stations
   B. the number & size of conductors being used
   C. transmission tower size and the power losses in the transmission
   D. All above

4. Why HVDC is a more efficient form of transmission than HVAC _____.
   A. The HVAC transmission lines require relatively larger spacing between the phase to ground & phase to phase conductors.
   B. The power losses in the HVDC transmission are significantly lower than HVAC
   C. HVAC needs to be taller & wider than HVDC

D. Using the HVDC transmission tower reduces the installation cost as compared to HVAC towers.

5. Break-even distance depends on _____.
   A. the type of transmission
   B. the transformer
   C. terminal station cost
   D. transmission line cost

Unit 21　Smart Grid

What is grid?

Concept of smart grid is quite in the news and market but the majority of the people actually don't know that what exactly are the things which make a grid smart[1]?

The term "Grid" refers to an "Electric Grid" basically describes a complete network which includes transmission lines, transformers, distribution substations and all accessories that are used for delivery of electricity from generation plants to home and commercial scale[2].

The very first grid was built in decade of 1885~1895, and with the passage of the time, number of grids kept on increasing that's why by now there are about more than 9200 grids all over the world which are providing about 1 million Megawatt power to the consumers.

As evolution has a direct relation with time, so for an efficient functionality of grid, digital technology has been introduced in grid system. This new digital technology enables two way communications which guarantee the direct link between utilities and all consumers.

What is a smart grid?

In simple words, a smart grid is an automation system between utility and consumers. This smart grid consists of advance digital system, automation, computer and control which make sure to perform a duplex "two way" communication between the power provider and load consumer.

In a typical electrical grid system, electricity providers only will know the power failure when a costumer call them. But in case of smart grid system, if electric supply fails, service providers will automatically respond to the affected area because the components of smart grid provide enough data i.e. from the power transformer, main transmission and distribution system and finally to the home supply system (you may say the utility meter)[3].

What things make a grid "Smart"?

According to the Department of Energy (United States), four types of advance technology will transform a typical electrical grid into smart grid which are as follow:
- Fully automated and Integrated two way communication among the overall components of an electric grid.
- Automatic control for power distribution, faults and repairs.
- Advance management panel, decision support software and mechanism.

- Accurate sensing and measurement technologies.

Upgraded technology of smart grids has well organized automation equipment and control system, whose response is accurate to meet the rapidly increasing demand for electricity. Time when these smart grids were not implemented, all utilities companies were bound to send their respective workers to take meter reading and acquire data related to consumer[4].

What does a smart grid do?

Smart grid performs lots of smart jobs. Some advantages of a smart grid are stated as follow:

1) Efficient transmission and distribution of electric power.
2) Quickly restore electric power after power failure due to faults.
3) Lower cost for operation, maintenance, management and electricity for both utilities and consumers.
4) Lower electric power tariff and rates due to reduced peak demand.
5) Provide better options of integration of renewable energy for self power generation systems.
6) Improve the security and protection.

Applications of a smart grid system(Fig.4.7)

Deployment of digital technology in smart grids ensures the reliability, efficiency and accessibility for consumers regarding all utilities which count towards the economic stability of the nation[5]. Right at the start of transition time it becomes perilous to execute testing, to improve the technology by upgradation, developing and maintaining standards on a standard threshold and also application of these efficient grids serves all these problems.

Fig.4.7 A smart grid system

Basic applications of smart grids are:
- They improve the adeptness of transmission lines.
- Quick recovery after any sudden breakage/disturbance in lines and feeders.

- Cost reduction.
- Reduction of peak demand.
- They possess the ability to be integrated with renewable energy sources on a large level which leads to sharing of load and reduction of load on a large scale.

Technical Words and Expressions

accessory	[əkˈsesəri]	n.	附件，配件
functionality	[ˌfʌŋkʃəˈnæləti]	n.	功能，功能性
duplex	[ˈduːˌpleks]	n.	双工
the Department of Energy			能源部
smart grid			智能电网
management panel			管理面板
power failure			电源故障
be bound to			一定要，决心
tariff	[ˈtærɪf]	n.	关税，（旅馆、饭店或服务公司的）价目表，收费表
electric power tariff			电价，上网电价
deployment	[dɪˈplɔɪmənt]	n.	部署
accessibility	[əkˌsesəˈbɪləti]	n.	可及性
renewable energy source			可再生能源
breakage	[ˈbreɪkɪdʒ]	n.	破损，损坏

Notes

[1] Concept of smart grid is quite in the news and market, but majority of the people actually dont know that what exactly are the things which make a grid smart?

译文：智能电网的概念在新闻和市场上都很流行，但大多数人实际上并不知道究竟是什么让电网变得智能？

注释：本句中含有一个宾语从句"that what exactly are the things which make a grid smart"，该宾语从句中又包含了一个定语从句"which make a grid smart"修饰"things"。

[2] The term "Grid" refers to an "Electric Grid" basically describes a complete network which includes transmission lines, transformers, distribution substations and all accessories that are used for delivery of electricity from generation plants to home and commercial scale.

译文：术语"电网"是指"电网"，基本上描述了一个完整的网络，包括输电线路、变压器、配电变电站以及用于将电力从发电厂输送到家庭和商业场所的所有配件。

注释：本句包含了一个定语从句"which includes ……"修饰"network"。该从句中又包含了一个"that"引导的定语从句"that are …… commercial scale"修饰"all accessories"。

[3] But in case of smart grid system, if electric supply fails, service providers will automatically respond to the affected area because the components of smart grid provide enough data i.e. from the power transformer, main transmission and distribution system and finally to the home supply system (you may say the utility meter).

译文：但在智能电网系统下，如果电力供应出现故障，服务提供商将自动应对受影响的区域，因为智能电网的组件提供了足够的数据，这些数据包括从电力变压器、主输配电系统，最后到家庭供电系统（你可以说是用户电能表）。

注释：该句为"if"引导的条件状语从句，主句为"service providers will automatically respond to ……"。主句中又含了一个"because"引导的原因状语从句。

[4] Time when these smart grids were not implemented all utilities companies were bound to send their respective workers to take meter reading and acquire data related to consumer.

译文：当这些智能电网没有实施时，所有电力公司都必须派遣各自的工人进行抄表来获取消费者的用电数据。

注释：该句中"when these smart grids were not implemented"为定义从句修饰"time"。"related to consumer"为过去分词作后置定语修饰"data"。

[5] Deployment of Digital Technology in smart grids ensures the reliability, efficiency and accessibility to the consumers regarding all utilities which count towards the economic stability of the nation.

译文：在智能电网中部署数字技术确保了所有用电公司的可靠性、效率和消费者的可访问性，这对国家的经济稳定至关重要。

注释：该句主语部分为"Deployment of Digital Technology in smart grids"，谓语部分为"ensures"，宾语部分为"the reliability, efficiency and accessibility to the consumers"。其中"which count towards the economic stability of the nation."为定语从句修饰"the reliability, efficiency and accessibility to the consumers"。

Exercises

Ⅰ. Mark the following statements with T(true) or F(false) according to the text.

1. The term "Grid" refer to an "Electric Grid". ( )
2. In a typical electrical grid system, electricity provider will know the power failure if electric supply fails. ( )
3. In a typical electrical grid system, we use accurate sensing and measurement technologies. ( )
4. Quick recovery after any sudden breakage/disturbance in lines and feeders is an advantage of a smart grid. ( )
5. Smart grids can reduce peak demand. ( )

Ⅱ. Complete the following sentences

1. The term "Grid" refer to an "_____" basically describes a complete network which includes _____, transformers, _____ all accessories that are used for delivery of _____

from generation plants to home and commercial scale.

2. This smart grid consist of advance _____, _____, computer and control which make sure to perform a duplex "two way" communication between the power provider and _____ consumer.

3. Deployment of digital technology in smart grids ensures the _____, _____ and accessibility to the consumers regarding all _____ which count towards the economic _____ of the nation.

4. Right at the start of _____ it become perilous to execute _____, to improve the technology by up gradation, developing and _____ standards on a standard threshold and also application of these efficient grids serve all these problems.

5. Smart grids possess the ability to be integrated with _____ on a large level which leads to sharing of _____ and reduction of load on large scale.

Reading 21 Integration of Renewable Energy with Grid System

Types of energy which exist infinitely and never run out completely are renewable forms of energy. Consider wind, coal, biomass, propane, uranium, water, sun, these are the sources that are naturally available to us, never run out and they were not formed.

Far in history concept of production of electricity from these sources stepped in the industry and with the passage of time proper production was started . According to a survey in 2009 this production was 19% of total 100% and this percentage has increased to about 49% according to latest survey in America in 2013. In European countries, these resources are getting encouraged to be merged in conventional distributive systems.

When we talk about grid systems, these systems allow us to allocate power everywhere in every application starting from a household need to commercial need of any level.

These resources are best alternative solution to meet all needs of electric load. Every part of the world is blessed with almost all means of renewable energy so it is most likely solution that if we can merge the production of renewable energy sources with our main distributing grid system.

These sources will take almost 70% of the load as long as water is flowing, wind is blowing, and sun is shining, or in other scenario if somehow these sources are temporaryily unavailable, our grid system can take load on momentary basis.

Now the question arises that how to merge these sources in our grid systems. The first main requirement is the balancing of the overall system and balancing is totally dependent on our needs. About half of the cost of total deployed system will be spent on this balancing system according to load and application requirements. But normal required equipments are as follow:
- Instrumentation/Parts.
- Requirements for connections of grids.
- Place for deployment, community and other specific requirements.

Balancing of overall grid system requires

- Power conditioning instrumentation.
- Safety requirement.
- Smart meters for efficient performance.

Once it has been decided that these sources are going to be connected with conventional grid system, the most simplest form is direct connection of produced current to the load. But if the storage is required, situation will change.

For storage batteries are required along with the charge controller. Standard diagram for storage(Fig.4.8) will look like.

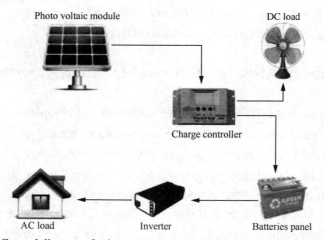

Fig.4.8　General diagram of solar power system with batteries storage, AC& DC load

Batteries are used as backups and are an effective mean of storage when connected to wind or photovoltaic electricity. Batteries (lead acid) are deep cycle batteries and they last for about 9～10 years if maintained properly.

Regulation of current to the battery is completely controlled by charge controller. Does the controller keep the battery charged full and also avoid the overcharging of battery? Moreover it also keeps check when extra current is drawn by load side and also prevents overcharging by deployment of shunt.

Then power conditioning is the most important thing that must be fulfilled. Energy produced by these sources is mostly in DC, so conversion of DC to AC is a must for AC transmission and distribution in the system.

- Matching of frequency.
- Matching of voltage.
- Matching of phase.
- Matching of constant power to oscillating.

Inverters are used to do this power conditioning. Costing of inverters is the next step which includes following things:

- Area and level of application.
- Quality of the electricity.
- Voltage of the incoming current.
- AC voltage required.
- Power requirements.
- Other inverter features such as meters and indicator lights.

Technical Words and Expressions

integration	n	整合，一体化，结合
biomass	n	生物质
propane	n	丙烷
uranium	n	铀
scenario	n	方案，设想
storage battery		蓄电池，蓄电池组
lead acid		铅酸，铅酸蓄电池
shunt	n	分流器
charge controller		充电控制器
overcharge	v	过度充电

Comprehension

1. Which following resources can be alternative solution to meet all needs of electric load? _____.

    A. biomass   B. wind   C. sun   D. all above

2. Renewable energy sources will take almost _____ of the load, if somehow these sources are temporary unavailable our grid system can take load on momentary basis.

    A. 19%   B. 49%   C. 70%   D. 100%

3. To merge renewable energy in our grid systems, equipments which we normal require are _____.

    A. instrumentation/parts

    B. requirements for connections of grid

    C. place for deployment, community and other specific requirements

    D. all above

4. Energy produced by renewable energy sources is mostly in _____.

    A. DC   B. AC   C. current   D. voltage

5. Which of the following things is not included in the costing of inverters? _____

    A. Area and level of Application   B. Quality of the electricity

    C. Power requirements   D. Matching of frequency

Chapter 5
Robot Technology

Chapter 5　Robot Technology

Unit 22　Types of Robot

Robots are designed to fulfill user needs. Robot types can be divided into:
- Manipulator robot, for example an arm robot;
- Wheeled robot;
- Walking robot;
- Humanoid robot;
- Aerial robot;
- Submarine robot.

A robot has these essential characteristics.
- Sensing: first of all your robot would have to be able to sense its surroundings. It would do this in ways that are not unsimilar to the way that you sense your surroundings. Giving your robot sensors: light sensors (eyes), touch and pressure sensors (hands), chemical sensors (nose), hearing and sonar sensors (ears), and taste sensors (tongue) will give your robot awareness of its environment.
- Movement: a robot needs to be able to move around its environment. Whether rolling on wheels, walking on legs or propelling by thrusters, a robot needs to be able to move. To count as a robot either the whole robot moves, like the Sojourner or just parts of the robot moves, like the Canada Arm.
- Energy: a robot needs to be able to power itself. A robot might be solar powered, electrically powered, battery powered. The way your robot gets its energy will depend on what your robot needs to do.
- Programmability: it can be programmed to accomplish a large variety of tasks. After being programmed, it operates automatically.
- Mechanical capability: enabling it to act on its environment rather than merely function as a data processing or computational device (a robot is a machine).
- Intelligence: a robot needs to be smart. A programmer is the person who gives the robot its 'smarts.' The robot will have to have some way to receive the program so that it knows what it is to do.

　　A manipulator is a device used to manipulate materials without direct contact. The applications were originally for dealing with radioactive or biohazardous materials, using robotic arms, or they were used in inaccessible places. In more recent developments they have been used in applications such as robotically assisted surgery and in space. It is an arm-like mechanism that consists of a series of segments, usually sliding or jointed, which grasp and move objects with a number of degrees of freedom.

　　Robot manipulators are created from a sequence of link and joint combinations. The links are the rigid members connecting the joints, or axes. The axes are the movable components of the robotic manipulator that cause relative motion between adjoining links. The mechanical joints used to construct the robotic arm manipulator consist of five principal types. Two of the joints are linear, in which the relative motion between adjacent links is nonrotational, and the other three of the joints are rotary types, in which the relative motion involves rotation between links (Fig.5.1).

The arm-and-body section of robotic manipulators is based on one of four configurations. Each of these anatomies provides a different work envelope and is suited for different applications.

- Gantry: these robots have linear joints and are mounted overhead. They are also called Cartesian and rectilinear robots.
- Cylindrical: named for the shape of its work envelope, cylindrical anatomy robots are fashioned from linear joints that connect to a rotary base joint.
- Polar: the base joint of a polar robot allows for twisting and the joints are a combination of rotary and linear types. The work space created by this configuration is spherical.
- Jointed-arm: this is the most popular industrial robotic configuration. The arm connects with a twisting joint, and the links within it are connected with rotary joints. It is also called an articulated robot.

Fig.5.1   4 DOF Manipulator

As the development of robot technology, the capability of the robot to "see" or vision based robots has been developed such as ASIMO, a humanoid robot created by Honda (Fig. 5.2). With a height of 130 centimeters and a weigh of 54 kilograms, the robot resembles the appearance of an astronaut with the ability fingers capable of handling egg. ASIMO can walk on two legs with a gait that resembles a human to a speed of 6 km/h. ASIMO was created at Honda's Research and Development Center in Wako Fundamental Technical Research Center in Japan. The model is now the eleventh version, since the commencement of the ASIMO project in 1986. According to Honda, ASIMO is an acronym for "Advanced Step in Innovative Mobility" (a big step in the innovative movement). This robot has a height of 130cm with a

Chapter 5 Robot Technology

total of 34 DOF and uses 51.8V LI-ION rechargeable and the ability of mechanical grip better.

The rapid development of robot technology has prompted the presence of intelligent robots capable of complement and assisted the work of man. The ability to develop robots capable of interacting today is very important, for example, the development of educational robot NAO from France and Darwin OP from Korea. In the latest development of robot vision are generally humanoid form, requires the Linux embedded module that can process images from the camera quickly. For example, Smart Humanoid robot package Ver. 2.0 for general-purpose robot soccer Which has the specification (Fig.5.3):

Fig.5.2   ASIMO robot

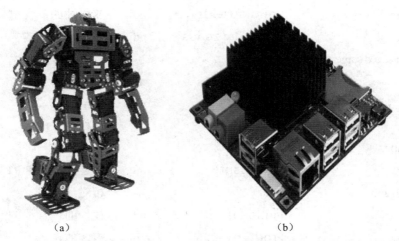

(a)   (b)

Fig.5.3   Smart Humanoid ver. 2.0 using embedded system and webcam based on LINUX Ubuntu

- CM-530 [Main Controller-ARM Cortex (32bits) with AX-12A (Robot Exclusive Actuator, Dynamixel].
- AX-18A (Robot Exclusive Actuator, Dynamixel).
- Gyro Sensor (2 Axis) Distance measurement system.
- RC-100A (Remote Controller).
- Rechargeable Battery (11V, Li-Po, 1000mA/PCM).
- Balance Battery Charger.
- Humanoid Aluminum frame full set.
- Gripper frame set.
- 1.7GHz Quad core ARM Cortex-A9 MP Core.
- 2GB Memory with Linux UBuntu.
- 6 x High speed USB2.0 Host port.
- 10/100Mbps Ethernet with RJ-45 LAN Jack.

Technical Words and Expressions

robot	['rəubɔt]	n.	机器人
manipulator	[mə'nɪpjuleɪtə(r)]	n.	操纵者，机械手；机器人操作器
arm robot			机械手机器人
submarine	[ˌsʌbmə'riːn]	adj.	水下的；海底的
sense	[sens]	vt.	感到；理解，领会；检测出
sonar	['səunɑː(r)]	n.	声呐
thruster	[θ'rʌstə]	n.	推进器
solar	['səulə(r)]	adj.	太阳的，日光的
Intelligence	[ɪn'telɪdʒəns]		智力，情报
radioactive	[ˌreɪdiəu'æktɪv]	adj.	放射性的；辐射
biohazardous	[biːəu'hæzədəs]	adj.	生物危害性的
robotically assisted surgery sliding			机器人辅助手术
sliding	['slaɪdɪŋ]	n.	滑块
joint	[dʒɔɪnt]	n.	关节
link	[lɪŋk]	n.	连杆
rigid member			刚性部件
adjacent	[ə'dʒeɪsnt]	adj.	邻近的，毗邻的
relative motion			相对运动
linear	['lɪniə(r)]	adj.	线性的，直线的
rotary	['rəutəri]	adj.	旋转的
anatomy	[ə'nætəmi]	n.	分解，解剖
gantry	['gæntri]	n.	（起重的）龙门架
cylindrical		adj.	圆筒形；圆柱形的
Cartesian	[kɑː'tiːziən]	adj.	笛卡尔坐标；直角坐标型
rectilinear	[ˌrektɪ'lɪniə(r)]	adj.	直线的，形成直线的；直线型
polar robot		adj.	极坐标型机器人
articulated robot			关节型机器人
embedded module			嵌入模块
gyro	['dʒaɪrəu]	n.	陀螺仪，回转仪
gripper	['grɪpə]	n.	抓手
quad core			四核处理器
MP core			多核处理器

Chapter 5 Robot Technology

 Notes

[1] The applications were originally for dealing with radioactive or biohazardous materials, using robotic arms, or they were used in inaccessible places.

译文：机器人最初是使用机器人手臂，来处理有辐射或化学危害的材料，或用在人类无法到达的区域。

注释：该句为 or 连接的并列句 The applications were originally for … , or they were used in …，其中 applications 和 they 指代"机器人"。"using robotic arms"部分为现在分词做状语。

[2] It is an arm-like mechanism that consists of a series of segments, usually sliding or jointed, which grasp and move objects with a number of degrees of freedom.

译文：它们是一种类似于人类手臂一样的机械装置，由滑块或关节这些部件组合而成，可以在一定的自由度下对物体进行抓取和移动。

注释：该句包含 that 引导的定义从句"that consists of … jointed"，which 引导的非限定性定义从句"which grasp and move objects with a number of degrees of freedom"对"arm-like mechanism"补充说明。

[3] Two of the joints are linear, in which the relative motion between adjacent links is nonrotational, and three are rotary types, in which the relative motion involves rotation between links.

译文：其中两个是线性的，与该类关节相链接的连杆作非旋转运动，还有三个是旋转的，与该类关节相链接的连杆作相对的旋转运动。

注释：该句为 and 连接的并列句 "Two of the joints are linear…" and "three are rotary types"，每个单句包含一个"in which"引导的定义从句。

[4] As the development of robot technology, the capability of the robot to "see" or vision based robot has been developed such as ASIMO, a humanoid robot created by Honda.

译文：随着机器人技术的发展，人们开发出具备了"看"的功能的机器人即视觉机器人如阿西莫，日本本田研制的类人机器人。

注释：句中"As … technology"为 as 引导的条件状语，翻译为"随着…"，主语为"the capability of … robot"，谓语部分为"has been developed"，"such as ASIMO… Honda"为插入语举例说明。

[5] The rapid development of robot technology has demanded the presence of intelligent robots capable of complement and assisted the work of man.

译文：机器人技术的快速发展催生了智能机器人的出现，这种机器人可以完成或辅助人们完成一些工作。

注释：句中"The rapid development of robot technology"为主语，"has demanded"为谓语，"the presence of intelligent robots"为宾语，"capable of complement and assisted the work of man"为定语从句，省略"which"，修饰"intelligent robots"。

189

Exercises

I. Mark the following statements with T (true) or F (false) according to the text.

1. The robot senses its surroundings in ways that are not similar to the way that you sense your surroundings. (   )
2. The robot has one way to receive the program. (   )
3. A manipulator is a device used to manipulate materials with direct contact. (   )
4. A polar robot is the most popular industrial robotic configuration. (   )
5. The ability to develop robots capable of interacting today is very important. (   )

II. Complete the following sentences.

1. Giving your robot sensors:_____(eyes), touch and _____ sensors (hands), chemical sensors (nose), hearing and _____ sensors (ears), and taste sensors (tongue) will give your robot awareness of its environment.

2. Robot manipulators are created from a sequence of _____ and _____ combinations. The links are the _____ connecting the joints or _____. The axes are the movable components of the robotic manipulator that cause _____ between adjoining links.

3. Jointed-Arm - This is the most popular _____ configuration. The arm connects with a twisting joint, and the links within I t are connected with _____ joints. It is also called an _____.

4. Polar——the base joint of a polar robot allows for _____ and the joints are a combination of _____ and _____ types. The work space created by this configuration is _____.

5. The rapid development of robot technology has demanded the presence of _____ robots capable of complement and _____ the work of man.

Reading 22 Arm Geometries

Generally, there are five configurations robots used in the industry, namely: Cartesian Robot, Robot Cylindrical, Spherical Robots, Articulated Robots (consist of revolute joint RRR), SCARA [Selectively Compliant Assembly Robot Arm (Fig.5.4)]. They are named for the shape of the volume that the manipulator can reach and orient the gripper into any position the work envelope. They all have their uses, but as will become apparent, some are better for use on robots than others. Some use all sliding motions, some use only pivoting joints, some use both. Pivoting joints are usually more robust than sliding joints, with careful design, sliding or extending can be used effectively for some types of tasks.

The Denavit-Hartenberg (DH) Convention is the accepted method of drawing robot arms in FBD's. There are only two motions a joint could make: translate and rotate. There are only three axes this could happen on: x, y, and z (out of plane). Below I will show a few robot arms, and then draw the Robot Arm Free Body Diagram (FBD). A cartesian coordinate robot (also called

Chapter 5 Robot Technology

linear robot) is an industrial robot whose three principal axes of control are linear (i.e. they move in a straight line rather than rotate) and are at right angles to each other. Cartesian coordinate robots with the horizontal member supported at both ends are sometimes called Gantry robots.

Example of manipulator for industry is KUKA KR 5 arc rounds off the range of KUKA robots at the lower end (Fig.5.5). Its payload of 5 kg makes it outstandingly well-suited to standard arc welding tasks. With its attractive price and compact dimensions, it is the ideal choice for your application too. Whether mounted on the floor or inverted overhead, the KR 5 arc always performs its tasks reliably.

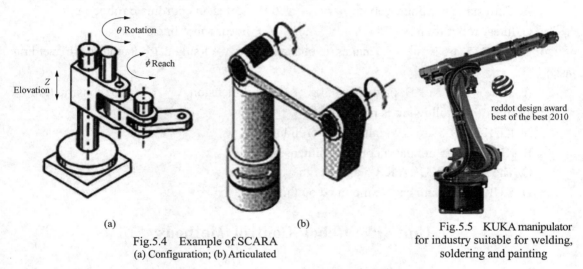

Fig.5.4   Example of SCARA
(a) Configuration; (b) Articulated

Fig.5.5   KUKA manipulator for industry suitable for welding, soldering and painting

Technical Words and Expressions

spherical	adj.	球形的，球面的
volume	n.	体积
Free Body Diagram (FBD)		自由物体受力图
straight line		直线
horizontal	adj.	水平的
arc welding		电弧焊
dimension	n.	尺寸; [复] 面积，范围; [数] 次元，度，维
mount	vt.	安装，架置

Comprehension

1. Which is not the configurations robots used in industry?
   A. Cartesian Robot
   B. Cylindrical Robot

C. Rotary Robot

D. Selectively Compliant Assembly Robot Arm

2. _____are usually more robust than _____ but, with careful design, sliding or extending can be used effectively for some types of tasks.

  A. Sliding joints pivoting joints   B. Saddle joints pivoting joints

  C. Sliding joints gliding joints   D. Pivoting joints sliding joints

3. A _____ is an industrial robot whose three principal axes of control are _____ and are at right angles to each other.

  A. Cartesian coordinate robot rotary   B. Cartesian coordinate robot curve

  C. linear robot rotary   D. linear robot linear

4. The KR 5 arc's _____ makes it outstandingly well-suited to standard arc welding tasks.

  A. price   B. payload of 5kg   C. dimension   D. construction

5. Which of the following is not right? _____.

  A. KUKA KR 5 arc is Manipulator for industry.

  B. KUKA KR 5 arc has a compact dimension.

  C. The price of KUKA KR 5 arc is high.

  D. KUKA KR 5 arc can be mounted on the floor.

Unit 23   Robot Control Methods

All robot control methods involve a computer, robot, and sensors. Next, we will introduce some basic programming and related sensors.

Lead-through programming

The human operator physically grabs the end-effector and shows the robot exactly what motions to make for a task, while the computer memorizes the motions (memorizing the joint positions, lengths and/or angles, to be played back during task execution).

Teach programming

Move robot to required task positions via teach pendant; computer memorizes these configurations and plays them back in robot motion sequence. The teach pendant is a controller box that allows the human operator to position the robot by manipulating the buttons on the box (Fig.5.6). This type of control is adequate for simple, non-intelligent tasks.

Fig.5.6   Microbot with teach Pendant

Off-line programming

Off-line programming is the use of computer software with realistic graphics to plan and program motions without the use of robot hardware (such as IGRIP).

Autonomous

Autonomous robots are controlled by computer, with sensor feedback, without human intervention. Computer control is required for intelligent robot control. In this type of control, the computer may send the robot pre-programmed positions and even manipulate the speed and direction of the robot as it moves, based on sensor feedback. The computer can also communicate with other devices to help guide the robot through its tasks.

Teleoperation

Teleoperation is human-directed motion, via a joystick. Special joysticks that allow the human operator to feel what the robot feels are called haptic interfaces.

Telerobotic

Telerobotic control is a combination of autonomous and teleoperation control of robot systems.

Robot sensors

Robots under computer control interact with a variety of sensors, which are small electronic or electro mechanical components that allow the robot to react to its environment. Some common sensors are described below.

Vision

A vision system has a computer-controlled camera that allows the robot to see its environment and adjust its motion accordingly. Used commonly in electronics assembly to place expensive circuit chips accurately through holes in the circuit boards. Note that the camera is actually under computer control and the computer sends the signals to the robot based upon what it sees.

Voice

Voice systems allow the control of the robots using voice commands. This is useful in training robots when the trainer has to manipulate other objects.

Tactile

Tactile sensors provide the robot with the ability to touch and feel. These sensors are used for measuring applications and interacting gently with the environment.

Force/Pressure

Force/pressure sensors provide the robot with a sense of the force being applied on the arm and the direction of the force. These sensors are used to help the robot auto-correct for misalignments, or to sense the distribution of loads on irregular geometry. Can also measure torques, or moments, which are forces acting through a distance. Can be used in conjunction with haptic interfaces to allow the human operator to feel what the robot is exerting on the environment during teleoperation tasks.

Proximity

Proximity sensors allow the robots to detect the presence of objects that are very close to the arm before the arm actually contacts the objects. These sensors are used to provide the robot with a method of collision avoidance.

Limit switches

Limit switches may be installed at end-of-motion areas in the workspace to automatically stop the robot or reverse its direction when a move out-of-bounds is attempted; again, used to avoid collisions.

Other sensors

- Encoder: measures angle;
- Potentiometer: measures angle or length;
- LVDT: measures length (linear variable displacement transducer);
- Strain gauge: measures deflection;
- Ultrasonic sensor: measures distance;
- Infrared sensor: measures distance;
- Light sensor: detects presence.

Technical Words and Expressions

lead-through programming			牵引规划
grab	[græb]	vt.	抓起，抓住
end-effector			末端执行器
joint	[ˈdʒɔint]	n.	关节；接合处
pendant	[ˈpendənt]	n.	配对物，匹配物，附属物，补充物
teach pendant			示教盒子
off-line			脱机，离线
autonomous	[ɔˈtɑnəməs]	adj.	自治的，自主的，自发的
teleoperation	[teləpəˈreɪʃn]	n.	远程操作
joystick	[ˈdʒɔiˌstik]	n.	操纵杆
haptic	[ˈhæptik]	adj.	触觉的
haptic interface			触觉界面
telerobotic	[ˌteliˈrɔbətik]	n.	遥控机器人

Chapter 5 Robot Technology

sensor	['sensə]	n.	传感器，灵敏元件
electro mechanical component			机电元件
vision	['viʒən]	n.	视觉
tactile	['tæktail]	adj.	触觉的，有触觉的，能触知的
misalignment	[misə'lainmənt]	n.	错位，调整不当，偏心度
geometry	[dʒi'ɔmitri]	n.	几何结构，几何图，几何形状
moment	['məumənt]	n.	力矩
proximity	[prɔk'simiti]	n.	接近
collision	[kə'liʒən]	n.	碰撞，冲突
avoidance	[ə'vɔidəns]	n.	避免
limit switch			限位开关
workspace			工作台，工作区
out-of-bound			界外
encoder	[in'kəudə]	n.	编码器
potentiometer	[patenʃi'ɔmitə]	n.	电位计，分压计
strain	[strein]	n.	紧张，拉紧
gauge	[gedʒ]	n.	标准度量，计量器
strain gauge			应变计，应变片
ultrasonic	['ʌltrə'sɔnik]	n.	超声波
infrared	['nfra'red]	adj.&n.	红外线的；红外线

 Notes

[1] Force/pressure sensors provide the robot with a sense of the force being applied on the arm and the direction of the force.

译文：力/压力传感器可以让机器人检测加在臂上的力以及力的方向。

注释：句中 a sense of the force 与 the direction of the force 为并列成分，being applied on the arm 作 the force 的后置定语。

[2] These sensors are used to help the robot auto-correct for misalignments, or to sense the distribution of loads on irregular geometry.

译文：这些传感器有助于机器人自动矫正误差或者检测不规则几何图形负载上的干扰。

注释：句中 to help the robot … 与 to sense the distribution …为并列状语。

 **Exercises**

Ⅰ. **Mark the following statements with T(true) or F(false) according to the text.**

1. Off-line programming is the use of computer hardware with realistic graphics to plan and program motions without the use of robot software.   (   )

2. Autonomous robots are controlled by computer, with sensor feedback, without human intervention. (   )

3. Telerobotic control is a combination of autonomous and teleoperation control of robot systems. (   )

4. Force sensors can't provide the robot with the ability to touch and feel. (   )

5. A voice system has a computer-controlled camera that allows the robot to see its environment and adjust its motion accordingly. (   )

II. Complete the following sentences.

1. The teach pendant is a _____ that allows the human operator to _____ the robot by manipulating the _____ on the box.

2. Off-line programming is the use of computer software with _____ to plan and program motions without the use of _____ (such as IGRIP).

3. Special joysticks that allow the _____ to feel what the robot feels are called _____.

4. Telerobotic control is a combination of _____ and _____ control of robot systems.

5. Limit switches may be installed at _____ in the workspace to automatically stop the robot or reverse its direction when a move _____ is attempted.

Reading 23 Robot Control

Off-line programming (OLP) is a robot programming method where the robot program is created independent from the actual robot cell. The robot program is then uploaded to the real industrial robot for execution. In off-line programming, the robot cell is represented through a graphical 3D model in a simulator. Nowadays OLP and robotics simulator tools help robot integrators create the optimal program paths for the robot to perform a specific task. Robot movements, reachability analysis, collision and near-miss detection and cycle time reporting can be included when simulating the robot program.

OLP does not interfere with production as the program for the robot is created outside the production process on an external computer. This method contradicts to the traditional on-line programming of industrial robots where the robot teach pendant is used for programming the robot manually.

The time for the adoption of new programs can be cut from weeks to a single day, enabling the robotization of short-run production.

Control

The mechanical structure of a robot must be controlled to perform tasks. The control of a robot involves three distinct phases – perception, processing, and action (robotic paradigms). Sensors give information about the environment or the robot itself (e.g. the position of its joints or its end effector). This information is then processed to be stored or transmitted and to calculate the appropriate signals to the actuators (motors) which move the mechanical.

The processing phase can range in complexity. At a reactive level, it may translate raw

Chapter 5   Robot Technology

sensor information directly into actuator commands. Sensor fusion may first be used to estimate parameters of interest (e.g. the position of the robot's gripper) from noisy sensor data. An immediate task (such as moving the gripper in a certain direction) is inferred from these estimates. Techniques from control theory convert the task into commands that drive the actuators.

At longer time scales or with more sophisticated tasks, the robot may need to build and reason with a "cognitive" model (Fig.5.7). Cognitive models try to represent the robot, the world, and how they interact. Pattern recognition and computer vision can be used to track objects. Mapping techniques can be used to build maps of the world. Finally, motion planning and other artificial intelligence techniques may be used to figure out how to act (Fig.5.8). For example, a planner may figure out how to achieve a task without hitting obstacles, falling over, etc.

Fig.5.7   Robot models

Fig.5.8   RuBot II robot

Autonomy levels

Control systems may also have varying levels of autonomy.
- Direct interaction is used for haptic or teleoperated devices, and the human has nearly complete control over the robot's motion.
- Operator-assist modes have the operator commanding medium-to-high-level tasks, with the robot automatically figuring out how to achieve them.
- An autonomous robot may go without human interaction for extended periods of time (Fig.5.9). Higher levels of autonomy do not necessarily require more complex cognitive capabilities. For example, robots in assembly plants are completely autonomous but operate in a fixed pattern.

Fig.5.9   TOPIO (a humanoid robot, played ping pong at Tokyo IREX 2009)

Another classification takes into account the interaction between human control and the machine motions.

- Teleoperation. A human controls each movement, each machine actuator change is specified by the operator.
- Supervisory. A human specifies general moves or position changes and the machine decides specific movements of its actuators.
- Task-level autonomy. The operator specifies only the task and the robot manages itself to complete it.
- Full autonomy. The machine will create and complete all its tasks without human interaction.

Technical Words and Expressions

optimal program path		优化程序路径
cycle time		周期，循环时间
contradict to		矛盾
marionette	n.	牵线木偶
cube	n.	立方体，立方
perception	n.	观念，洞察力，认识能力
paradigm	n.	范例，示范，典范
cognitive	adj.	认识的，认知的，有认识力的
autonomy	n.	自治，自治权，自主

Comprehension

1. Which of the following is not right? _____.

　　A. Off-line programming (OLP) is a robot programming method where the robot program is created dependent from the actual robot cell.

　　B. Nowadays OLP and robotics simulator tools help robot integrators create the optimal program paths for the robot to perform a specific task.

　　C. The control of a robot involves three distinct phases – perception, processing, and action (robotic paradigms).

　　D. Direct interaction is used for haptic or teleoperated devices, and the human has nearly complete control over the robot's motion.

2. Off-line programming (OLP) is a robot programming method where the robot program is created _____ from the actual robot cell.

　　A. dependent　　B. dependent　　C. derived　　D. produced

3. The control of a robot involves three distinct phases - perception, _____, and action (robotic paradigms).

A. converting   B. deriving   C. processing   D. process

4. An autonomous robot may go without human interaction for _____ periods of time .

A. extended   B. short   C. temporary   D. certain

Unit 24   Embedded Systems for Robot

The robotics system requires adequate processor capabilities such as the ability of the processor speed, memory and I/O facilities. Fig. 5.10 is a block diagram of an intelligent robotics that can be built by beginners.

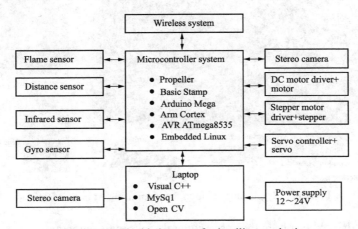

Fig.5.10   Embedded system for intelligent robotics

From the picture above(Fig.5.10), the point is you can use a variety of microprocessor/ microcontroller to make the robot as smart as possible. You may use the standard minimum systems such as Propeller, AVR, Basic Stamp, and Arm Cortex with extraordinary abilities. All inputs received by the sensors will be processed by the microcontroller. Then through the programs that we have made microprocessor / microcontroller will take action to the actuator such as a robot arm and the robot legs or wheels. Wireless technology used for the purposes of the above can transmit data or receive commands remotely. While the PC / Laptop is used to program and perform computational processes data / images with high speed, because it is not able to be done by a standard microcontroller. To provide power supply to the robots, we can use dry battery or solar cell. For the purposes of the experiment, can be used as a standard microcontroller for main robot controller as shown below using Arduino Mega.

The Fig.5.11 shows that the standard microcontroller technologies such as AVR, Adriano or Propeller and Arm Cortex, can be used as the main controller of mobile robots. Technology sensors and actuators can be handled well using a microcontroller with I^2C capability for data communication between the microcontrollers with a serial devices others. Some considerations in choosing the right microcontroller for the robot is the number of I/O, ADC capability, and signal processing features, RAM and Flash program memory. In a complex robot that requires a variety of sensors and large input the number, often takes more than one controller, which uses the principle of master and slave. In this model there is a 1 piece main controller which functions to coordinate the slave microcontroller.

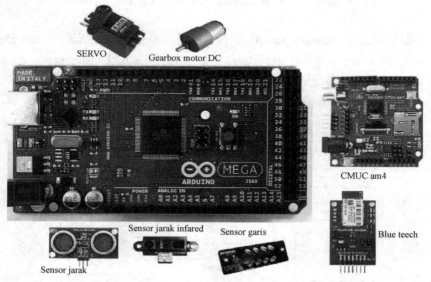

Fig.5.11 Single chip solution for robot using Arduino Mega

In general, to drive the robots there are several techniques such as:
- Single wheel drive, which is only one front wheel that can move to the right and to the left of the steering.
- Differential drive, where 2 wheels at the back to adjust the direction of motion of the robot.
- Synchronous drive, which can drive a 3 wheeled robot.
- Pivot drive, It is composed of a four wheeled chassis and a platform that can be raised or lowered. The wheels are driven by a motor for translation motion in a straight line.
- Wheels tank, Tracked robot uses wheels tank (Fig.5.12).
- Ackermaan steering, where the motion of the robot is controlled by the 2 front wheels and 2 rear wheels.
- Omni directional drives, where the motion of the robot can be controlled by 3 or 4 wheeled system that can rotate in any direction, so that the orientation of the robot remains. Omniwheel is useful because the orientation of the robot is fixed with the standard wheel angle $\alpha_1 = 0°$, $\alpha_2 = 120°$ and $\alpha_3 = 240°$. Global frame $[x, y]$ represents robot's environment and the location of robot can be represented as (x, y, θ). The global velocity of robot can be represented as $(\dot{x}, \dot{y}, \dot{\theta})$.

Chapter 5 Robot Technology

Fig.5.12 Tracked robot

🔌 Technical Words and Expressions

processor	['prəusesə(r)]	n.	计算机的中央处理器
microprocessor /microcontroller	[ˌmaɪkrəu'prəusesə(r)]	n.	微处理器；微控制器
extraordinary	[ɪk'strɔːdnri]	adj	非凡的；特别的；非常奇特的
minimum system			单片机最小系统
actuator	['æktʃueɪtə]	n.	[电脑]执行机构；[电]（电磁铁）螺线管；[机]驱动器
laptop	['læptɑp]	n.	便携式电脑；笔记本
dry battery			干电池
solar cell			太阳能电池
flash program memory			闪存程序存储器
steering	['stɪərɪŋ]	n.	转向装置；掌舵
single wheel drive			单轮驱动
differential drive			差分驱动
synchronous drive			同步驱动
chassis	['ʃæsi]	n.	（车辆的）底盘；（飞机的）起落架
rear wheel			后轮
pivot drive			轴驱动
omni wheel			万向轮
straight line			直线
rear wheel			后轮

201

Notes

[1] All inputs received by the sensors will be processed by the microcontroller.

译文：传感器检测到的所有输入信号由微控制器处理。

注释：句中主语为"All inputs"，"received by the sensors"为后置定语修饰"All input"。

[2] Wireless technology used for the purposes of the above can transmit data or receive commands remotely.

译文：上述所采用的无线电技术能传输数据或远程接收指令。

注释：该句中"used for the purposes of the above"作主语 Wireless technology 的后置定语。

[3] In a complex robot that requires a variety of sensors and large input the number, often takes more than one controller, which uses the principle of master and slave.

译文：复杂的机器人中经常需要多种传感器及大量的输入信号，因而复杂机器人需要配备多个控制器，多个控制器之间按照主从关系工作。

注释："that requires a variety …… the number"作句子的主语。谓语为 takes ，"which uses the …… slave" which 引导的非限制性定语从句修饰"controller"。

[4] Omni directional drives, where the motion of the robot can be controlled 3 or 4 wheeled system that can rotate in any direction, so that the orientation of the robot remains.

译文：万向轮驱动，机器人的运动控制由 3 或 4 轮系统控制，可在任何方向上旋转，因此机器人的方向任意可调。

注释：该句为"so that"引导的结果状语从句。主句为"Omni directional drives, where the …… wheel system that can rotate in any direction"其中包含了一个"where"引导的定语从句和"that"引导的定语从句修饰"Omni directional drives"。

Exercises

I. Mark the following statements with T(true) or F(false) according to the text.

1. Through the programs that we have made microprocessor / microcontroller will take action to the input. ( )

2. A standard microcontroller can program and perform computational processes data / images with high speed. ( )

3. A complex robot requires a only one controller. ( )

4. Ackermaan steering, where the motion of the robot is controlled by the 2 front wheels and 2 rear wheels. ( )

5. Synchronous drive, which can drive a 3 wheeled robot. ( )

II. Complete the following sentences.

1. The robotics system requires adequate _____ capabilities such as the ability of the processor speed, memory and _____ facilities.

2. All inputs are received by the _____ will be processed by the microcontroller.

3. Some considerations in choosing the right microcontroller for the robot is the number of I / O, ADC capability, and _____ features, RAM and _____.

4. Pivot drive, It is composed of a four wheeled chassis and a _____ that can be raised or lowered. The wheels are driven by a motor for translation motion in a _____.

5. Omni wheel is useful because the _____ of the robot is fix with the standard wheel angle $\alpha_1 = 0°, \alpha_2 = 120°$ and $\alpha_3 = 240°$.

Reading 24 Gear

Modern robotics needs excellent gears. A good understanding of how gears affect parameters such as torque and velocity are very important. Gears work on the principle of mechanical advantage. This means that by using different gear diameters, you can exchange between rotational (or translation) velocity and torque.

Introduction of gears

With gears, you will exchange the high velocity with a better torque. This exchange happens with a very simple equation that you can calculate:

Torque_Old * Velocity_Old = Torque_New * Velocity_New

Torque_Old and Velocity_Old can be found simply by looking up the datasheet of your motor. Then what you need to do is put a desired torque or velocity on the right hand side of the equation. So for example, suppose your motor outputs 3 lb-in torque at 2000rps according to the datasheet, but you only want 300rps. This is what your equation will look like:

3 lb-in * 2000rps = Torque_New * 300rps

Then you can then determine that your new torque will be 20 lb-in. The gearing ratio is the value at which you change your velocity and torque. Again, it has a very simple equation. The gearing ratio is just a fraction which you multiple your velocity and torque by. Suppose your gearing ratio is 3/1. This would mean you would multiple your torque by 3 and your velocity by theinverse or 1/3 .

Example: Torque_Old = 10 lb-in, Velocity_Old = 100rps

Gearing ratio = 2/3; Torque * 2/3 = 6.7 lb-in; Velocity * 3/2 = 150rps

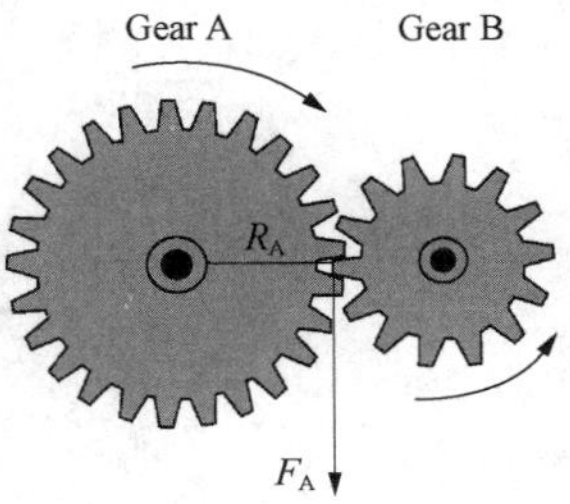

If you wanted a simple gearing ratio of say 2 to 1, you would use two gears, one being twice as big as the other (Fig.5.13). It isn't really the size as much as the diameter ratio of the two gears. If the diameter of one gear is 3 times bigger than the other gear, you would get a 3/1 (or 1/3) gearing ratio. You can easily figure out the ratio by hand measuring the diameter of the gears you are using.

Fig.5.13 Torque that generates to rotates gear B equal to $F_A R_A$

For a much more accurate way to calculate the gearing ratio, calculate the ratio of teeth on the gears. If one gear has 28 teeth and the other has 13, you would have a (28/13=2.15 or 13/28=0.46) 2.15 or 0.46 gearing ratio. I will go into this later, but this is why worm gears have such high gearing ratios. In a worm gear setup, one gear always has a single tooth, while the other has a guaranteed huge ratio. Counting teeth will always give you the most exact ratio.

Unfortunately, by using gears, you lower your input to output power efficiency. This is due to obvious things such as friction, misalignment of pressure angles, lubrication, gear backlash (spacing between meshed gear teeth between two gears) and angular momentum, etc. For example, suppose you use two spur gears, you would typically expect efficiency to be around 90%. To calculate , multiply that number by your Velocity_New and Torque_New to get your true output velocity and torque.

Gearing ratio = 2/3;
Torque×2/3=6.7 lb-in;
Velocity×3/2=150rps;
True torque=6.7×0.9=6 lb-in;
True velocity=150×0.9=135rps.

Types of gears

Some types of gears have high efficiencies, or high gearing ratios, or work at different angles. for example, often manufacturers will give you expected efficiencies in the datasheets for their gears. Remember, wear and lubrication will also dramatically affect gear efficiencies. Spur gears are the most commonly used gears due to their simplicity and the fact that they have the highest possible efficiency of all gear types (Fig.5.14). Not recommend for very high loads as gear teeth can break more easily.

Two gears with a chain can be considered as three separate gears (Fig.5.15). Since there is an odd number, the rotation direction is the same. They operate basically like spur gears, but due to increased contact area there is increased friction (hence lower efficiency). Lubrication is highly recommended.

Fig.5.14 Spur gears (with ~90 % efficiency)

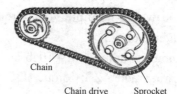

Fig.5.15 Sprocket gears with chains (with ~80% efficiency)

Worm gears have a very high gearing ratio. To mathematically calculate, consider the worm gear as a single tooth (Fig.5.16). Another advantage to the worm gear is that it is not back-drivable. What this means is only your motor can rotate the main gear, so things like gravity or counter forces will not cause any rotation. This is good say if you have a robot arm holding something heavy, and you don't want to waste power on holding torque.

Rack and Pinion is the type of gearing found in steering systems (Fig.5.17). This gearing is great if you want to convert rotational motion into translational. Mathematically, use radius = 1 for the straight 'gear'.

Fig.5.16 Worm gears (with ~70% efficiency)

Fig.5.17 Rack and pinion (with ~90% efficiency)

Technical Words and Expressions

gear	n.	齿轮
parameter	n.	参数
torque	n.	转矩
velocity	n.	速度
diameter	n.	半径
rotational velocity		转速
datasheet	n.	数据表
gearing ratio		齿轮比
fraction	n.	[数]分数
friction	n.	摩擦，摩擦力
misalignment	n.	未对准，角误差
lubrication	n.	润滑，加油，油润
gear backlash		齿隙，齿轮侧隙，回差
angular momentum		角动量
spur gear		直齿圆柱齿轮
sprocket gear		链轮
worm gear		涡轮

Comprehension

1. If your motor outputs 3 lb-in torque at 2000rps according to the datasheet, but you only want 400rps , your new torque will be _____.

A. 15 lb-in   B. 20lb-in   C. 25lb-in   D. 30 lb-in

2. Suppose your gearing ratio is 4/1, Torque_Old = 10 lb-in, Velocity_Old = 100rps, your new torque will be _____ and your new velocity will be _____.

    A. 25 lb-in   40 rps   B. 2.5 lb-in   400 rps

    C. 40 lb-in   25 rps   D. 400 lb-in   2.5 rps

3. If you wanted a simple gearing ratio of say 2 to 1, one gear has _____ teeth and the other has _____ teeth.

    A. 28   13   B. 13   28   C. 20   40   D. 20   1

4. If your gearing ratio is 4/1 and the efficiency of gear is around 90%, Torque_Old = 10 lb-in, Velocity_Old = 100rps, your output torque will be _____ and your output velocity will be _____.

    A. 25 lb-in   40 rps   B.36 lb-in   22.5 rps

    C. 40 lb-in   25 rps   D. 400 lb-in   22.5 rps

5. _____ have a very high gearing ratio.

    A. Spur gears   B. Sprocket gears

    C. Worm gears   D. Rack and Pinion

Unit 25 Robot Vision

There are several important terms in the robot vision interconnected, including computer vision, machine vision and robot vision. Computer vision is the most important technology in the future in the development of interactive robots. Computer vision is a field of knowledge that focuses on the field of artificial intelligence and systems associated with acquisition and image processing. Machine vision is implemented process technology for image-based automatic inspection, process control, and guiding robots in various industrial and domestic applications. Robot vision is the knowledge about the application of computer vision in the robot. The robot needs vision information to decide what action is to be performed. The application is currently in robot vision are as robot navigation aids, search for the desired object, and other environmental inspection. Vision on the robot becomes very important because it received more detailed information than just the proximity sensor or other sensors. For example, the robot is able to recognize whether the detected object is a person's face or not. Furthermore, an advanced vision system on the robot makes the robot can distinguish a face accurately (Face recognition system using PCA method, LDA and others). The processing of the input image from the camera to have meaning for the robots is known as visual perception, starting from image acquisition, image preprocessing to obtain the desired image and noise-free, for example, feature extraction to interpretation as shown in Fig.5.18.

An example of intelligent robotics is a humanoid robot HOAP-1 with stereo vision for navigation system (Fig.5.19). HOAP-1 is a commercial humanoid robot from Fujitsu Automation Ltd. and Fujitsu Laboratories Ltd. for behavior research. In the vision sub-system of HOAP-1, the depth map generator calculate depth map image from stereo images. The path planning sub-system generate a path from the current position to the given goal position while avoid obstacles.

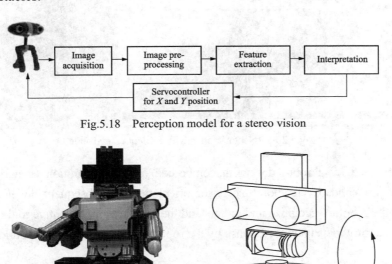

Fig.5.18   Perception model for a stereo vision

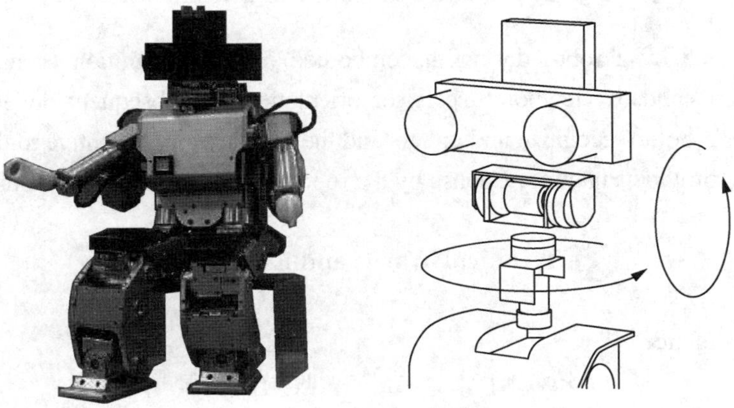

Fig.5.19   Example of vision-based navigation system for humanoid robot HOAP-1
(a) Humanoid robot; (b) 2DOF neck module with yaw and pitch axis

Another example is a telepresence robot developed by author as shown in Fig.5.20. The test was conducted by running Microsoft IIS and Google Application Engine on the laptop. When the servers were ready, Master Controller, implemented by using a laptop, opened the application through web browser that support WebRTC and entered 192.168.1.101 which was the address of both servers to open it[3]. This is not a problem because the servers were running on different ports. After the connections were established, Master controller then received image and sound stream from the robot and sent back image and sound from Master Controller web camera to the robot. Experiments of intelligent telepresence robot had been tested by navigating the robot to staff person and to avoid obstacles in the office. Face tracking and recognition based on eigenspaces with 3 images every person had been used and a databases of the images had been developed. The robot was controlled using integrated web application (ASP.Net and WebRTC) from Master Control. With a high speed Internet connection, simulated using wireless router that had speed around 1Mbps, the result of video conferencing was noticeable smooth.

Fig.5.20 Teleconference with a client using robot

Images collected by a robot during the embodied object recognition scenario often capture objects from a non-standard viewpoint, scale, or orientation. In subsequent development, artificial intelligence for the robot to recognize and understand the human voice, attentive to the various motion listener and able to provide a natural response by the robot are challenge ahead to build future robots.

Technical Words and Expressions

artificial intelligence			人工智能
acquisition	[ˌækwɪˈzɪʃn]	n.	收购,获得
inspection	[ɪnˈspekʃn]	n.	检验,检查,视察
process control			过程控制
navigation	[ˌnævɪˈgeɪʃn]	n.	航行(学),航海(术),导航
proximity	[prɑkˈsɪməti]	n.	接近,邻近
recognition	[ˌrekəgˈnɪʃn]	n.	认识,识别
visual perception			视觉检测
image acquisition			图像获取
image preprocessing			图像处理
feature extraction			特征提取
face tracking			面部跟踪
sub-system			子系统
depth map			深度图
master controller			主控制器
browser	[ˈbraʊzə(r)]	n.	浏览程序,浏览器
simulated	[ˈsɪmjuleɪtɪd]	adj.	仿造的,模仿的
router	[ˈruːtə(r)]	n.	路由器(传送信息的专用网络智能设备)
capture	[ˈkæptʃə(r)]	n.	[计] 捕捉

Chapter 5　Robot Technology

| scenario | [səˈnɑːriəu] | n. | （行动的）方案，剧情概要 |
| eigenspace | [ˈaigən,speis] | n. | 特征空间，本征空间 |

💡 Notes

[1] Computer vision is a field of knowledge that focuses on the field of artificial intelligence and systems associated with acquisition and image processing.

译文：计算机视觉所涉及的知识领域主要集中在人工智能与数据采集和图形处理系统。

注释：该句的主干为"Computer vision is a field of knowledge and systems"，其中"that focuses on the field of artificial intelligence"为定语从句修饰"knowledge"，"associated with acquisition and image processing"为过去分词做定语修饰"systems"。

[2] The processing of the input image from the camera to have meaning for the robots is known as visual perception, starting from image acquisition, image preprocessing to obtain the desired image and noise-free, for example, feature extraction to interpretation as shown in Fig. 5.18.

译文：处理自摄像机的输入图像对机器人的意义是视觉感知，从图像获取图像，再进行图像预处理从而获得预期的图像并且无噪声，例如，图 5.18 所示的特征提取到转换。

注释：该句的主干为"The processing is known as visual perception"，"of the input image from the camera to have meaning for the robots"为"processing"的后置定语，"starting from …to… to interpretation"为现在分词做宾语补足语。

[3] When the servers were ready, Master Controller, implemented by using a laptop, opened the application through web browser that support WebRTC and entered 192.168.1.101 which was the address of both servers to open it.

译文：当服务器准备就绪，使用便携式笔记本，主控制器通过支持 WebRTC 的网络浏览器打开应用软件，并输入 192.168.1.101，这是打开应用程序的两台服务器的地址。

注释："When the servers were ready"为时间状语从句，主句主干为"to open it（Master Controller）"，其中"implemented…, opened … and entered …servers"为过去分词做方式状语，"that support WebRTC"做定语从句修饰"browser"，"which was the address of both servers"定语从句修饰"192.168.1.101"。

[4] The robot was controlled using integrated web application (ASP.Net and WebRTC) from Master Control.

译文：机器人由主控制器上的集成网站应用软件来控制（如 ASP.Net 和 WebRTC）。

注释：该句主干为"he robot was controlled"，"using … Master Control"为现在分词做状语。

⚠ Exercises

Ⅰ. Mark the following statements with T(true) or F(false) according to the text.

1. Computer vision is the most important technology in the future in the development of interactive robots.　　　　　　　　　　　　　　　　　　　　　　　　　　　　（　）

2. Computer vision is the knowledge about the application of Robot vision in the robot. (   )

3. Vision on the robot received more detailed information than just the proximity sensor or other sensors. (   )

4. In the vision sub-system of HOAP-1, the depth map generator calculate depth map image from stereo images. (   )

5. Images collected by a robot during the embodied object recognition scenario often capture objects from a standard viewpoint, scale, or orientation. (   )

II. Complete the following sentences.

1. The application is currently in robot vision are as robot _____ aids, search for the desired object, and other environmental _____.

2. Furthermore, an advanced vision system on the robot makes the robot can _____ a face (Face _____ system using PCA method, LDA and others).

3. The processing of the input image from the camera to have meaning for the robots known as _____, starting from image _____, image _____ to obtain the desired image and _____, for example, _____ to interpretation.

4. An example of _____ is a humanoid robot HOAP-1 with _____ vision for navigation system.

5. _____ and recognition based on eigenspaces with 3 images every person had been used and a _____ of the images had been developed.

Reading 25 Sensors for Intelligent Robot

Ultrasonic distance sensor: PING

PING ultrasonic sensor provides an easy method of distance measurement. This sensor is perfect for any number of applications that require you to perform measurements between moving or stationary objects. Interfacing to a microcontroller is a snap. A single I/O pin is used to trigger an ultrasonic burst (well above human hearing) and then "listen" for the echo return pulse.

The sensor measures the time required for the echo return, and returns this value to the microcontroller as a variable-width pulse via the same I/O pin. The PING sensor works by transmitting an ultrasonic (well above human hearing range) burst and providing an output pulse that corresponds to the time required for the burst echo to return to the sensor. By measuring the echo pulse width, the distance to target can easily be calculated.

Key features:
- Provides precise, non-contact distance measurements within a 2cm to 3m.
- Range for robotics application.
- Ultrasonic measurements work in any lighting condition, making this a Good choice to

supplement infrared object detectors.
- Simple pulse in/pulse out communication requires just one I/O pin.
- Burst indicator LED shows measurement in progress.
- 3-pin header makes it easy to connect to a development board, directly or with an extension cable, no soldering required.

The PING sensor detects objects by emitting a short ultrasonic burst and then "listening" for the echo. Under control of a host microcontroller (trigger pulse), the sensor emits a short 40 kHz (ultrasonic) burst. This burst travels through the air, hits an object and then bounces back to the sensor. The PING sensor provides an output pulse to the host that will terminate when the echo is detected, hence the width of this pulse corresponds to the distance to the target (Fig.5.21).

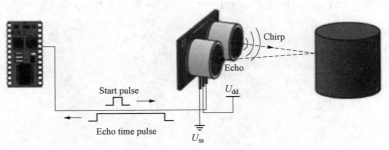

Fig.5.21 The basic principle of ultrasonic distance sensor

Technical Words and Expressions

ultrasonic	adj.	超声的，超音波的
stationary	adj.	不动的，固定的，静止的
trigger	vt.	引发，触发
pulse	n.	脉冲
ultrasonic burst	n.	超声波脉冲
precise	adj.	精密，精确的，清晰的
detector	n.	检测器，探测器
terminate	vt. &n.	结束，使终结，终点站

Comprehension

1. Ultrasonic sensor can be used to measure _____?
   A. voice   B. color   C. distance   D. weight
2. Ultrasonic sensor knows the distance to target through _____.
   A. the echo pulse amplitude   B. the echo pulse phase
   C. the echo pulse frequency   D. the echo pulse width
3. Which distance ultrasonic sensor can't measure? _____

A. 2cm   B. 20cm   C. 200cm   D. 2000cm

4. Which of the following states about The PING sensor is wrong? _____

   A. It can't work in night.

   B. It can measure distance of object without contacting.

   C. It can measure distance of object with high precision.

   D. All above.

5. Which of the following states about the ultrasonic sensor detects objects is wrong? _____

   A. The sensor emits a short ultrasonic burst.

   B. Ultrasonic burst travels through the light and hits the target

   C. The ultrasonic burst travels back to the sensor.

   D. The PING sensor provides an output pulse to host whose width is the distance to the target.

附录1 电气专业课程词汇中英文对照

电工电子技术	Electrical Technology & Electronic Engineering
电工基础理论	Fundamental Theory of Electrical Engineering
电工技术基础	Fundamentals of Electrotechnics
电工实习	Electrical Engineering Practice
电工学	Electrical Engineering
电路理论	Theory of Circuit
电路理论基础	Fundamental Theory of Circuit
电路理论实验	Experiments in Theory of Circuit
电子技术课程设计	Course Exercise in Electronic Technology
模拟电子技术	Analog Electronic Technique
模拟电子线路	Analog Electronic Circuitry
数字电子技术	Digital Electronic Technique
数字电子线路	Digital Electrical Circuitry
工业电子学	Industry Electronics
电子辅助设计	Electronic Aided Design
工程制图	Graphing of Engineering
金工实习	Machinery Practice
电机学	Electrical Motor
电力电子学	Power Electronics
电机学及控制电机	Electrical Machinery & Control Motor
电机与拖动	Electrical Machinery & Towage
传感器与检测技术	Sensors & Testing Technology
传感器原理	Fundamentals of Sensors
自动检测技术与仪表	Automatic Measurement Technique & Meter
电子测试技术	Electronic Testing Technology
单片机原理	Fundamentals of Mono-Chip Computers
单片机原理及应用	Fundamentals of Mono-Chip Computers & Applications
电气控制技术	Electrical Control Technology
电气控制与PLC应用	Electrical Control and Application of PLC
机床电气控制	Electrical Control of Machinery Tools
工厂供电	Factory Electricity Supply
楼宇自动化	Building Automation
综合布线技术	Structured Cabling Technology
计算机控制系统	Computer Control System
现场总线与工业以太网	Fieldbus & Industrial Ethernet

机电一体化	Mechanical & Electrical Integration
自动控制理论	Theory of Automatic Control
自动控制系统	Automatic Control System
自动控制原理	Principle of Automatic Control
过程控制	Process Control
组态控制技术	Industry Configuration Technology
微机原理及接口技术	Principle & Interface Technique of Microcomputer
微机原理及应用	Principle & Application of Microcomputer
机器人技术基础	Fundamentals of Robot Techniques
机器人控制技术	Robot Control Technology
专业外语	Specialized English
毕业实习	Graduation Practice
毕业设计	Graduation Design

附录2 论文英文摘要的书写

英文摘要是对所写文章主要内容的精炼概括。通常要求的论文摘要的字数不等,一般为200~500字。而国际刊物要求所刊登的论文摘要的字数通常是100~200字。摘要的位置一般放在一篇文章的最前面,内容上涵盖全文,并直接点明全旨。语言上要求尽量简练。摘要通常多采用第三人称撰写。摘要通常有标题、内容、关键字组成。现在我们来看一下各部分的书写格式。

一、标题(Title)

英文题名应以短语为主要形式,尤以名词短语最常见,即题名基本上由一个或几个名词加上其前置和(或)后置定语构成。短语型题名要确定好中心词,再进行前后修饰。各个词的顺序很重要,词序不当会导致表达不准。英文题名的大小写可以采用以下实词首字母大写,虚词小写。

二、摘要(Abstract)

以"Abstract"开头,接着由论文研究的问题或目的、过程和方法、结果三部分组成。

(1) 说明研究的问题或目的,通常用一般现在时或一般过去时。常用句型有以下几种。

1) This paper/article describes/reports/presents/discusses… 本文叙述/报告/介绍/讨论……

2) In this paper, we/the authors present… 本文介绍……

3) A study of…is reported 本文报告……的研究

4) In this paper…is (are) presented. 本文介绍……

(2) 说明方法和结果,一般用一般过去时、第三人称的被动语态表达。常用句型有:

1) Using… (technique), we studied… 我们用……(技术)研究了……

2) …was (were) measured using… (我们)用……测定了……

3) …was (were) analyzed (reviewed) by… (我们)用……分析(回顾)了……

4) The results showed (demonstrated) that… 结果表明……

5) It was found/observed that… (我们)发现/观察……

6) …demonstrated a significant improvement in… ……方面呈明显改善

(3) 结论通常用一般现在时、被动语态表达,也可用主动语态表达。常用句型有以下几种。

1) These results suggest that… 结果提示……

2) These findings indicate that… 这些发现表明……

3) the data (study) show/ demonstrate/illustrate that… 数据(研究)表明/证明/说明……

4) We conclude/ suggest/ believe that… 我们的结论是/建议/认为……

三、关键字(Keywords)

科技论文的关键词应从其题名、层次标题和正文中选出,应能反映论文的主题概念。英文摘要中列出关键词是为了满足国内外情报检索机构编制索引和二次文献及便于读者选读,所以关键词的选择须满足下列原则:

(1) 能代表论文主要内容;

(2) 每篇论文应选关键词3~8个,以保证论文的检出率;

(3) 关键词应选用主题词，以便于检索机构的检索；
(4) 英文摘要中的关键词与中文关键词应保持意思和顺序的一致性。

四、范例

Variational Bayesian Blind Deconvolution Using a Total Variation Prior

Abstract: In this paper, we present novel algorithms for total variation (TV) based blind deconvolution and parameter estimation utilizing a variational framework. Using a hierarchical Bayesian model, the unknown image, blur, and hyperparameters for the image, blur, and noise priors are estimated simultaneously. A variational inference approach is utilized so that approximations of the posterior distributions of the unknowns are obtained, thus providing a measure of the uncertainty of the estimates. Experimental results demonstrate that the proposed approaches provide higher restoration performance than non-TV-based methods without any assumptions about the unknown hyperparameters.

Keywords: total variation　blind deconvolution　variational framework　Bayesian model

参 考 文 献

[1] AND I V, SIWY Z S. Nanofluidic Diode [J]. Nano Letters, 2015, 7(3):2-6.

[2] BLAIR T H. Variable Frequency Drive Systems [M] // Energy Production Systems Engineering. John Wiley & Sons, Inc. 2016:55-58.

[3] EMILIO M D P. Operational Amplifier [M] // Microelectronics. Springer International Publishing, 2016:112-125.

[4] KORNUTA T. Robot Control System Design Exemplified by Multi-Camera Visual Servoing [J]. Journal of Intelligent and Robotic Systems, 2015, 77(3):499-523.

[5] Kunegis, Jérôme. Handbook of Network Analysis [KONECT -- the Koblenz Network Collection] [J]. Computer Science, 2014(2):1343-1350.

[6] LI H, YANG L. The Application of Fuzzy Control in PLC Temperature Control System Based on OPC Technology [C] // Seventh International Symposium on Computational Intelligence & Design. IEEE, 2015:68-75.

[7] OZEKI M. Development of the Control System with Versatile PLCs for the Long-pulse Negative ion Source [J]. Social Cognitive & Affective Neuroscience, 2015, 7(1):11-22.

[8] Shanmugasundram R, Zakariah K M, Yadaiah N. Implementation and Performance Analysis of Digital Controllers for Brushless DC Motor Drives [J]. IEEE/ASME Transactions on Mechatronics, 2014, 19(1):213-224.

[9] Wang, Yinshun. Fundamental Elements of Applied Superconductivity in Electrical Engineering (Wang/Fundamental) // Case Study of Superconductivity Applications in Power System-HTS Cable [J]. 2013:389-420.

[10] Yongchang Z, Haitao Y. Model Predictive Flux Control for Induction Motor Drives [J]. Proceedings of the CSEE, 2015, 35(3):719-726.

[11] ZHANG C, HE X F. A Wind Energy Powered Wireless Temperature Sensor Node [J]. Sensors, 2015, 15(3):20-31.